T0075203

Proven Impossible

In mathematics, it simply is not true that "you can't prove a negative." Many revolutionary impossibility theorems reveal profound properties of logic, computation, fairness, and the universe and form the mathematical background of new technologies and Nobel prizes. But to fully appreciate these theorems and their impact on mathematics and beyond, you must understand their proofs.

This book is the first to present complete proofs of these theorems for a broad, lay audience. It fully develops the simplest rigorous proofs found in the literature, reworked to contain less jargon and notation, and more background, intuition, examples, explanations, and exercises. Amazingly, all of the proofs in this book involve only arithmetic and basic logic – and are elementary, starting only from first principles and definitions.

Very little background knowledge is required, and no specialized mathematical training – all you need is the discipline to follow logical arguments and a pen in your hand.

DAN GUSFIELD is Distinguished Professor Emeritus, and the former department chair, in the Department of Computer Science at University of California, Davis. He is Fellow of the ACM, the IEEE, and the ISCB. His previous books are *The Stable Marriage Problem* (1989, co-authored with Rob Irving), *Strings, Trees and Sequences* (1997), *ReCombinatorics* (2014), and *Integer Linear Programming in Computational and Systems Biology* (2019). As this book reflects, his teaching emphasized mathematical rigor as well as accessibility and clarity. He produced over 100 hours of video lectures on a wide range of topics, now viewed over a million times on the web.

Proven Impossible

Elementary Proofs of Profound Impossibility from Arrow, Bell, Chaitin, Gödel, Turing and More

DAN GUSFIELD
University of California, Davis

Shaftesbury Road, Cambridge CB2 8EA, United Kingdom

One Liberty Plaza, 20th Floor, New York, NY 10006, USA

477 Williamstown Road, Port Melbourne, VIC 3207, Australia

314–321, 3rd Floor, Plot 3, Splendor Forum, Jasola District Centre, New Delhi – 110025, India

103 Penang Road, #05–06/07, Visioncrest Commercial, Singapore 238467

Cambridge University Press is part of Cambridge University Press & Assessment, a department of the University of Cambridge.

We share the University's mission to contribute to society through the pursuit of education, learning and research at the highest international levels of excellence.

www.cambridge.org
Information on this title: www.cambridge.org/9781009349505

DOI: 10.1017/9781009349451

First published 2024

A catalogue record for this publication is available from the British Library

A Cataloging-in-Publication data record for this book is available from the Library of Congress

ISBN 978-1-009-34950-5 Hardback
ISBN 978-1-009-34949-9 Paperback

This book is dedicated to Sylvia Spengler, and to the memory of Frank Olken.

Although my involvement with Sylvia and Frank is completely unrelated to the topics in this book (and much more related to previous books of mine), I wanted to express my gratitude to them through this dedication.

Sylvia Spengler, through her own research, her vision and advocacy, her role as the interim director of the Human Genome Center at Lawrence Berkeley National Labs (before NIH became active in human genomics), and her work as a long-time program director at the National Science Foundation, together with the assistance and enthusiasm of Frank Olken, has been a major force advocating for federal support of computational biology research in the United States. She has had, and continues to have, a huge impact on the intersection of computer science and genomic biology, and on the development of a community of researchers at that intersection. Sylvia consistently encouraged my work throughout my career in computational biology. Without her advocacy and support, the academic computational biology community would not have developed as strongly as it has. She rightfully deserves our deepest appreciation.

Frank was a bibliophile and voracious reader with very broad interests, a lover of science, and an engaging conversationalist who always had interesting ideas and suggestions. He was an enthusiastic booster. I think he would have enjoyed this book. He died too young.

Contents

Preface

The Impossible brings confusion and sorrow
Certainty of the impossible brings harmony and joy

Writing this book has been a labor of love – it was never work. I always enjoyed researching the topics in the book, learning much more about some of them than I had known before. Each topic demonstrates the power of pure reason and the brilliance of the people who did the reasoning. And the reading, writing, and arithmetic that I did for the book made me feel as though I was having personal communications with those brilliant minds.

I hope you will enjoy reading (and working through) this book as much as I enjoyed writing it, and that you will find each impossibility result as inspiring as I did.

P.S. If I'm ever asked what I did during the COVID isolation and pandemic of 2020–2022, and how (or if) I maintained my sanity, I will point to this book. It kept me inside and COVID-free until almost the end of 2022 (COVID positive on November 30, 2022).

Acknowledgments

I want to thank Chip Martel, Lenny Pitt, and Steve Rawlinson for reading portions of the book in draft form and making very helpful suggestions. I thank Bruno Nachtergaele for researching the early use of the term "hidden variables" and Professor Geoffrey Pullum for the use of his poetry. I also thank the three anonymous reviewers of the draft submitted to Cambridge University Press, for their very helpful and detailed suggestions, most of which were incorporated in the final version of the book. And, I want to thank the editors at Cambridge University Press for their enthusiastic support of this project.

1

Yes You *Can* Prove a Negative!

1.1 Introduction to Impossibility

I have always reacted negatively when I hear the claim that "You can't prove a negative", because in theoretical computer science and mathematics, we do it all the time. Many famous negative results go back thousands of years. The beautiful ancient Greek proof that the square root of two cannot be written as the ratio of two integers (in other words, that $\sqrt{2}$ is *irrational*) is perhaps the best known example. Mathematics is full of such "negative" proofs.[1]

But in the 1900s, much deeper, more sweeping and "more negative" types of negative theorems were developed. These showed very unintuitive, profound, shocking, revolutionary, and even dangerous[2] types of impossibility, with implications far beyond their technical statements.

The first two of these sweeping, profound impossibility theorems were due to Kurt Gödel in 1931. Those two theorems (roughly) showed (a) the

[1] A recent book [63] and [89] explains the impossibility of solving four famous mathematical problems from antiquity. Some were open for several thousand years before being shown to be impossible. The book [65] focuses on a different famous problem from antiquity: whether there are formulas analogous to the quadratic formula (that solves equations with largest exponent equal to two), but for equations with exponents larger than two. After many failures, it was finally proven in the 1800s that there is no such analogous formula for equations with exponents of five (or larger), but there is a formula for equations where the largest exponent is three, and a different formula where the largest exponent is four. Those formulas are much more complex than the quadratic formula, which is taught to most high-school students.

[2] In Stalin's Soviet Union, some of these impossibility results were dangerous for anyone to study and teach. Stalin stated that "if there is a passionate desire to do so, every goal can be reached, every obstacle can be overcome" [53] – everything was achievable in the Utopian world of New Soviet Man. Moreover, "Stalin believed that science should serve the state. Pure research was not merely an indulgence - it was counter-productive. It was tantamount to wrecking [sabotage against the state]," and "theory must be refashioned to be of immediate service to the revolution" [53]. Many scientists whose work didn't fit with Stalin's views were sent to gulags or just disappeared. While this mostly affected biologists and physicists, logicians also knew to be careful – asserting impossibility was "counter-revolutionary", and dangerous.

impossibility of finding an axiomatic theory of arithmetic in which all and only true statements about arithmetic can be derived and (b) the impossibility of finding a consistent theory of arithmetic that can prove its own consistency.[3]

At about the same time, related work by Alfred Tarski showed the impossibility of *defining truth* about arithmetic, in the language of arithmetic itself. Then followed the work of Alan Turing, who showed that many problems are impossible to *solve* by (always correct) algorithms, and hence by computer programs. Turing's theorems can also be used to prove Gödel-like theorems. Several decades later, Gregory Chaitin proved an impossibility theorem that can also be used to prove Gödel-like theorems but is considered to be even more "devastating" and "limiting" for mathematics than are Gödel's and Turing's theorems.

In a different domain, Bell's impossibility theorem and inequalities have had enormous impact in physics, philosophy, and more recently on technological developments such as quantum computing and quantum cryptography. In fact, it is Bell's theorem and inequalities that form the mathematical basis for the experiments done by three physicists who won the 2022 Nobel prize in Physics. Some people consider Bell's work to be the most impactful development in all of science. And yet, and despite many lay expositions, Bell's theorem and inequalities are not well known outside of physics, philosophy, and now computer science.[4] And even inside physics, many physicists [47] have not mastered or studied the technical details of Bell's theorem and inequalities.[5]

In the areas of Economics and Political Science, the impossibility-of-fair-elections theorem, proved by Kenneth Arrow in 1950, initiated a subarea of Economics called *social-welfare theory*. It led to many other impossibility theorems concerning voting, the most compelling (to me) being the Gibbard–Satterthwaite theorem (which we prove before proving Arrow's theorem) concerning the impossibility of devising fair election mechanisms where *deception* is never rewarded. Arrow's impossibility theorem has wide "name recognition" outside of professional circles, since Arrow shared a Nobel prize in Economics, in large part for his seminal contributions to social-welfare theory.

[3] I know this second statement is vague – it will be more fully explained in Chapter 9.

[4] Although, in 1993, David Mermin [71] wrote (somewhat derisively I think): "[Bell's theorem] is widely known not only among physicists, but also to philosophers, journalists, mystics, novelists, and poets."

[5] Scott Aaronson, a computer scientist, has written that he has had physicists ask him to explain Bell's inequalities. [1] p. 109

Consistent with the axiomatic framework of Arrow, Jon Kleinberg more recently proposed three axioms he asserted should be obeyed by good *data clustering* algorithms. Clustering is a somewhat ill-defined task, of huge importance in many data-intense areas, most recently in the growing field of *machine learning* and the emerging field of *data science*. Kleinberg showed that there is no algorithm that can simultaneously obey those three axioms. This impossibility also holds for other, more natural, axioms, but as we will see, those results are less robust.

1.2 What These Theorems and Proofs Have in Common

All of these (and other) deep, profound (even revolutionary) impossibility theorems have three things in common:

(a) They are the product of pure reason.
(b) They were originally (and some still are) considered too hard for nonmathematicians or nonspecialists to fully grasp, that is, to follow actual, rigorous, correct *proofs*.[6]
And most important:
(c) We are no longer bound by Statement (*b*)!

Statement (*b*) is less relevant than it once was, because simpler proofs, generalizations, and/or simpler expositions are now known that *should* allow

[6] For example, a recent book on Bell's theorem [47] states: "I have made no attempt to show the actual theorem Bell had proved. Rather, I have talked about it only in general terms. The reason is that his combination of mathematical quantities is something of a complicated mess, and the proof that they obey his restriction is more complicated still." A similar statement: "Bell's theorem is a mathematical construct which as such is indecipherable to the non-mathematician" appears in [113].

Concerning Gödel's incompleteness [23]: "At the time of its discovery, ... its proof was considered to be extremely difficult and recondite" [23], and in the major popularization of Gödel's proof [77]: "The details of Gödel's proofs in his epoch-making paper are too difficult to follow without considerable mathematical training." A related quote from [39]: "... even some accomplished mathematicians (for example ...) had difficulty grasping the proof."

Similarly, Arrow's original proofs are extremely hard to follow. His initial 18-page paper, [4], presented his general impossibility theorem but only proved the special case of two individuals (voters) rather than an unlimited number. As stated in [95]: "He uses a mathematical wave of the hand on the generalizability of proofs involving societies with two individuals." Arrow's original paper [4] only states: "... the results stated in this paper hold for any number of individuals." But the proof of the general theorem, given in Arrow's book [5] a year later, is very different from his proof of the special case. He provided a different, but still long and difficult, proof, in a second edition [6], 12 years later.

nonmathematicians and nonspecialists to understand and appreciate these profound theorems and proofs.[7]

1.3 This Book Builds on These Improvements

Even though the newer proofs are simpler, existing *expositions* of these proofs are still largely written in terse, mathematical language aimed at "mathematically mature" professionals. And unfortunately,

> ... in the finished product of mathematics it is not uncommon to find that all signs of intuition have been removed. [21]

That situation is the motivation for this book.

This book takes advantage of, and builds on, the newer, simpler proofs and generalizations, with the goal of bringing diverse, rigorous, impossibility proofs to a broad, *nonprofessional, lay* audience. The book requires little background knowledge, and no specialized mathematical training – but it does require work.

To achieve the goal of exposing fully rigorous, yet accessible proofs, I searched the various literatures (and the Internet) for the most accessible proofs and expositions, and then built on those to reduce the jargon and notational burden, and to provide more background, intuition (when possible), examples, and explanations, in slower-paced presentations. Sometimes just making the text less dense, adding headers and summaries, and spacing out equations makes the presentation easier to understand.

For example, the 2005 proof that I follow in discussing Arrow's general impossibility theorem was published in a research journal [41] in typical

[7] For example, the quote in the previous footnote from [113] was noted by David Mermin in [68], followed by the statement: "a view I hope the rest of this article will dispel." I think Mermin largely realized that hope.

With respect to Gödel's theorems, the quote in the previous footnote from [23] continues: "With the passage of time the situation has been reversed. A great many different proofs of Gödel's theorem are now known, and the result is now considered easy to prove and almost obvious." Well, I don't agree that its proof is easy or obvious, and many of the different proofs share the same high-level ideas (see [60] for a list and discussion of several proofs), but there are certainly now vastly easier and more understandable proofs and expositions than Gödel's original one.

And continuing the earlier quote from [39]: "Today, as in the case of other intellectual advances, both the subject and our understanding of it have developed to the point where the proof is not at all considered difficult. ... proofs have become streamlined and generalized."

With regard to Arrow's theorem, there have been huge improvements. There are now *single-page* proofs of Arrow's general theorem. And even Kleinberg's original proof on clustering, which is only a few years old, has since been simplified and extended.

terse-math style, and occupies *less than one page*, including a figure.[8] My slower, more explanatory retelling of that proof expands the exposition to eight pages, with seven figures.

1.4 Why Proofs?

I believe that understanding (or at least following) a rigorous *proof* of a revolutionary theorem is the best way to absorb and appreciate the revolution; to appreciate the power of pure reason and the beauty of mathematics; to understand its impact beyond mathematics (on philosophy, science, the mind, technology, and society); and to have a justified sense of personal accomplishment, especially when you have heard that only a few highly trained people can possibly understand these proofs.

Understanding the proof is also the best way to understand any continuing *controversies* and misuses of a theorem.[9] All of the theorems in this book are to some extent controversial and/or commonly misunderstood – particularly the theorems of Gödel and Bell. There are numerous and egregious examples of these theorems being misunderstood and/or abused, and there are books and articles that discuss these misuses, abuses, and controversies. It seems to me almost impossible (maybe even provably impossible!) to fully understand the controversies, and certainly impossible to come to some conclusion about them yourself, if you don't understand the technical details of the theorems and proofs. For example, the larger meaning of Bell's theorem continues to be debated in physics and philosophy, interconnected with debates about the larger meaning of quantum mechanics itself. But, the math underneath Bell's theorem (presented well) is *not* difficult, and understanding the math brings one much closer to facing the mystery and controversies of its *meanings*.

Emphasizing these points:

> Everything is vague to a degree you do not realise till you have tried to make it precise. Bertrand Russell [94]

> Often in mathematics, the proof of a result turns out to be more important than the result itself. [29]

[8] The same journal even published essentially (in my view) the same proof 10 years later by a different author, where the only selling point of the second paper was that it had even *less* explanation than the first one. That kind of terseness is prized in professional mathematical writing but is absolutely contrary to what I am trying to do in this book.

[9] The misunderstanding (by funding agencies and researchers) of an impossibility theorem concerning *perceptrons* is believed to have stalled (later successful) research into *artificial neural networks* (an area of enormous current importance) for more than a decade [63].

Finally, understanding the proofs (or even just the general scope and nature of the proofs) will help you spot (common) *nonsensical* statements about famous theorems. For example, after the chapters on Gödel's theorems, you should have no trouble seeing that the following is total nonsense, unrelated to what Gödel's theorems actually concern:

> Basically, Gödel's Theorems prove the Doctrine of Original Sin, ... and that there is a future eternity. Gödel's Theorems mean that, in the human complex, things will go wrong and there will always be a "defect" of sorts about which forgiveness and corrective action will be needed. [80]

1.5 OK, Proofs Are Important, but Why Another Book?

It is the act of a madman to pursue impossibilities.

–Marcus Aurelius

Most of the topics in this book are well known, and some are the focus of many books and articles. So, why write another book on these topics?

1.5.1 Because

Because almost all[10] of the theorems in this book, despite how difficult their original proofs and expositions were, now have elementary, accessible, yet rigorous mathematical proofs. However, most expositions for *lay* audiences do not attempt rigorous *proofs*, and sometimes do not even attempt rigorous *statements* of the theorems. They just try to give the *gist*, or they just *talk about* the theorems, and maybe speculate on their larger impact. Worse, some expositions use loose, incorrect statements about the theorems to support ideological, religious, political, or philosophical viewpoints.[11]

For example, most discussions of Gödel's theorems fall into one of two types: either (1) they emphasize perceived philosophical "meanings" of the theorems, and maybe sketch some of the ideas of the proofs, usually relating Gödel's theorems and proofs to riddles and paradoxes, but they do not attempt to present rigorous, complete proofs or (2) they do present rigorous proofs, but in the traditional style of mathematical logic, with all of its heavy, weird

[10] I wrote "all" not "almost all" first, when I had decided not to discuss Gödel's Second Incompleteness Theorem – since I do not know an accessible, complete proof of it. But, then I changed my mind.

[11] And some writers doing this have the appropriate background and intellectual firepower to know better!

notation, difficult definitions, and technical issues that reflect Gödel's original motivation, proofs, and extensions.

Many people are frustrated by these two extreme types of expositions and want a complete, rigorous proof of incompleteness that they can understand, even if it doesn't exactly prove the same incompleteness that Gödel did. Fortunately, after Gödel published his seminal paper in 1931, people (read Turing and others) realized that somewhat weaker versions of Gödel's first incompleteness theorem have much simpler proofs and expositions. And, these variants of Gödel's first incompleteness theorem still have (in the opinion of many) most of the moral and philosophical force of Gödel's first incompleteness theorem. This book develops several such proofs.

Similarly, there are many lay-audience expositions of Bell's impossibility theorem and inequalities, but most only talk *about it* and/or its perceived philosophical implications.[12]

1.5.2 In Contrast

This book is as much about mathematical proofs as it is about impossibility.

My goal in writing this book is to make simpler, full, rigorous proofs of some of the most profound reasoning in human history accessible to a very broad, lay audience. All of the (full) proofs I discuss in this book involve only *elementary arithmetic and logic* – no advanced math or advanced logical manipulations are required. The book has complete, self-contained, mathematically rigorous proofs for most of the theorems that are discussed.

And, the proofs are "elementary" (in a different sense), meaning that they derive from basic *definitions* and *first principles*. And, although the theorems come from diverse application areas (physics, economics, logic, computer science, and machine learning), none of the proofs assume advanced knowledge of any application.

> Sometimes I've believed as many as six impossible things before breakfast. –
> Lewis Carroll

[12] With one great exception: The first half of the paper [68] by David Mermin, although originally published in a professional physics journal, is mostly (in my opinion) accessible to a larger lay audience. And, it *does* present a rigorous proof of Bell's impossibility theorem, although that isn't always appreciated. OK, Mermin's exposition doesn't have a lot of fancy mathematical symbols; it isn't written in the most obscure, terse way possible; and it doesn't even have the words "Theorem" or "Proof", but that doesn't keep it from being a truly rigorous mathematical argument. Much of Chapters 2 and 3 of this book are strongly influenced by Mermin's expositions, particularly [68], [70], and [72].

I believe that anyone with a high-school knowledge of mathematics, and the discipline to follow logical arguments (actively, with pen in hand), can understand all of the full proofs given in this book. The only background that some readers might not have is a very basic understanding that a computer program consists of a series of *textual* statements.[13] But, no actual programming knowledge is required, and I have a short Appendix where I discuss what one needs to know about computer programs.

1.5.3 Other Impossibilities

There are several well-developed impossibility theorems that I had initially hoped to include, but in the end I ran out of time and space. I recommend these topics and sources to interested readers. There are many impossibility proofs in the area of *distributed computation* and the *web*. That is an area summarized in the paper: "A hundred impossibility proofs for distributed computing" [67] and a whole book: *Impossibility Results for Distributed Computing* [7]. Also, see [92, 61] for an interesting impossibility theorem in the area of blockchains and Bitcoin.

1.6 How to Read This Book

All expositions are aimed at a general, lay audience, with no specialized background in any of the topics. But that doesn't mean the book is *light* reading. It takes work to fully understand the proofs. The most important thing is to read *actively* with pen in hand, and then to restate, re-justify, and explain in your own words the theorems and proofs. An even better way to learn is to try to *teach* the theorems and proofs to someone else – or pretend to.[14]

In several chapters that have long-ish derivations, I have broken up the expositions with fairly straightforward *Review questions*, to encourage readers to read actively and assure themselves that they are following the ideas in the chapter. Exercises at the end of each chapter are a bit more challenging than the review questions, and some are open-ended discussion questions.

The book is roughly divided into three parts that can be read *independently*. Chapters 2 and 3 form one part, and they should be read in order. Chapters 4

[13] As opposed, say, to the science-fiction image of a computer, probably named *HAL*, with some kind of mechanical (or green bubbling, gooey) brain that just spits out answers in a mysterious, condescending manner.

[14] It took me 40 years of teaching college students before I realized this and was able to give a helpful answer to the recurrent question of how best to study and learn – learn by teaching, if only to your invisible friend!

and 5 form a second part, but can be read separately. Chapters 6–9 form the third part, although the technical material in Chapter 8 is self-contained and can be read separately. Chapter 9 could be read separately, but I don't advise it since it is probably the mathematically hardest chapter in the book[15] – at least read Chapter 6 first for a warm-up. The Appendix is a short discussion on what a computer *program* is, for anyone who has never seen one, making the crucial point that a computer program is a textual document.

[15] The mathematically easiest chapter is Chapter 2. The hardest-*looking* proofs are in Chapter 4, but that is only because they are the longest. Each step of those proofs is intended to be fairly simple, and I have included many *review questions* in that chapter.

2

Bell's Impossibility Theorem(s)

Or, how can we know if the moon is really there when no one looks?

2.1 Bell, Einstein, and All That

Arguably, the most profound and impactful impossibility theorem in this book (and maybe in the whole Universe) is Bell's theorem (also known as Bell's inequality).[1,2] Quoting from a 1985 paper by David Mermin [68]:

> ... We read, for example, in The Dancing Wu Li Masters that:
>
> "Some physicists are convinced that [Bell's theorem] is the most important single work, perhaps, in the history of physics."[3]
>
> And indeed, Henry Stapp, a particle theorist at Berkeley, writes that:
>
> "Bell's theorem is the most profound discovery of science."

These statements may be too strong, but there is no question that Bell's theorems are deep and profound, and expose the fact that some parts of the physical world are very different from what we normally encounter in our everyday lives. Furthermore, the statements above were made *before* anyone realized that the revolutionary *impossibility* proved by Bell could lead to revolutionary *possibility:*

[1] Actually, Bell showed the first inequality, but after that, many other related inequalities were derived. Most people call any of these theorems a "Bell theorem" or a "Bell inequality", and we will do so also.

[2] In physics, impossibility theorems are often called "No-Go theorems".

[3] In quotations, words that are enclosed by square brackets have been added to the original quote, to more clearly connect the quote to the exposition.

> Bell's theorem ... prompted physicists to explore how quantum mechanics might enable tasks unimaginable in a classical world. "The quantum revolution that's happening now, and all these quantum technologies – that's 100% thanks to Bell's theorem,"[4] says Krister Shalm, a quantum physicist at the National Institute of Standards and Technology. [19]

And, amazingly, the derivation of Bell's theorem only requires a bit of *simple arithmetic* and *simple logic*.

What are Bell's theorems about? Bell's theorems concern what *kinds of theories* might be able to explain strange physical phenomena that were first predicted and then observed in many real experiments.[5] The phenomena occur in the *quantum world* and are predicted by *quantum mechanics*, which is the formal, mathematical framework for reasoning and making predictions about the quantum world.[6]

A very little about Quantum Mechanics Quantum mechanics, developed in the 1920s by Niels Bohr, Werner Heisenberg, and Erwin Schrödinger, is a mathematical/physical theory that is now the basis for *almost all* of modern physics. It is considered the most accurate theory of physical phenomena that has ever been developed. It is particularly effective in explaining and predicting physical phenomena in the sub-microscopic world (e.g., atoms, electrons, photons, neutrinos, quarks, etc.). This sub-microscopic, even subatomic, world is sometimes called the *quantum world*.

In contrast to quantum mechanics, the physical theory that applies to larger objects, and to the typical physical phenomena that we encounter in our everyday world, is part of what is now called *classical physics*. That is essentially the physics built on the physical laws developed by Isaac Newton.[7] Classical physics is what is typically taught in high school or in a first college course on physics.

[4] An example of this is reflected in the title of a recent paper in *Nature* on secure cryptography: "Experimental Quantum Key Distribution Certified by Bell's Theorem" [76].

[5] The 2022 Nobel prize in Physics was awarded to three physicists, Alain Aspect, John Clauser, and Anton Zeilinger, who independently designed and conducted the first such experiments. The award citation reads: "For experiments with entangled photons, establishing the violation of Bell inequalities and pioneering quantum information science."

[6] Don't worry. You don't need any previous knowledge of quantum mechanics, and we won't discuss any *hard-core* quantum mechanical mathematics.

[7] Einstein's relativity theory (special relativity was published in 1905, and general relativity in 1915) is also considered part of classical physics, but we won't be discussing relativity theory at all.

Objects in the quantum world can exhibit behaviors that are very different from what we encounter in the world of classical physics. Thus, those behaviors are unintuitive to us, and seem weird when we try to explain them using concepts and language developed for objects and behaviors seen in our normal world. Classical physics does not explain those behaviors,[8] but quantum mechanics does.

Further, based on our experiences, people make certain *assumptions* about the physical world that are so intuitive and natural that they are often not even stated. In fact, most people just take these assumptions as obvious truths, and not assumptions at all.[9] Such (often unstated) assumptions are built into classical physics.

But, these unstated, intuitive assumptions come into sharp focus in quantum mechanics, because quantum theory asserts that many of these assumptions *do not hold* in the quantum world. And, such assumptions are at the heart of what bothered Einstein about quantum mechanics.

Later (in Section 2.3.4), we will explicitly discuss what bothered Einstein about quantum theory, what kind of alternate theories he thought must exist, and how those concerns relate to Bell's Impossibility theorem(s). For now, we just say that Einstein insisted that a more "reasonable" theory, based more on classical physics, should exist. Quoting from [18]:

> ... what Einstein was trying to do, [was] find a theory underneath quantum mechanics that is more intuitive in our world.

We will call such a theory an *Einsteinian* theory, and describe it more completely later.

Bell's question

> Is there an Einsteinian theory (based on classical physics) that can make the *same* predictions that quantum mechanics makes? Or is such a theory *impossible*?

Essentially, Bell [13, 14] proved the latter: Such theories are *impossible*.

And, it's elementary Illustrating one of the major themes of this book, Bell's theorem and inequality (and its relatives) can be established with only *elementary reasoning* and *simple arithmetic*. No deep mathematics or complex logic

[8] **Spoiler Alert:** What Bell's theorem does is change "does not explain" to "cannot explain".

[9] Experiments with babies show that many of these assumptions about our physical world are "hard-wired", built into our brains even without much life experience, and certainly without any explicit instruction.

is needed. The genius of Bell was in imagining that such theorems were possible, when brilliant and driven people (such as Einstein, Bohr, and Schrödinger) overlooked them and seemingly did not even imagine that such results were possible, for thirty years.[10]

> We emphasize that the premises of Bell are those of [Einstein and associates in [33]] ... but he [Bell] revealed an unexpected consequence of those premises. [46]

And, no quantum mechanical math is needed The reasoning involved in Bell's inequality does *not* involve any mathematical reasoning about quantum mechanics. It only involves mathematical reasoning about our normal, classical physics, "common sense" world. But, the *relevance* and the *motivation* for Bell's work *do* involve quantum mechanics: quantum mechanics predicts certain phenomena (which are now widely observed) that *violate* Bell's inequalities, showing that quantum mechanics *must* go outside of our "common sense notions of how the world works". That conclusion is what is formally, and forcefully, established by Bell's theorem(s).

2.2 Is Impossibility Possible or Impossible?

Before proceeding, we need to ask and answer: How can one ever *prove* that it is *impossible* to find an Einsteinian theory, that is, an explanation, for quantum phenomena that obeys Einstein's assumptions? How can we even *specify* what kind of theories we are considering?

There may be some specific theories that one could analyze to see if they properly explain known quantum phenomena, and for each such theory, we might show that it does not. Results of that type can be very valuable, but we want to do more than just analyze *specific* proposed theories. We want to prove *sweeping* statements about *all conceivable* theories of the type that Einstein insisted on. That is what we want to do.

Well, we can't do it!

Why not?

The problem is that *we don't know what we don't know.* We don't know what "all conceivable" theories are. And, this is an issue that arises in almost all impossibility proofs.

[10] Well, actually this is not quite correct. In 1932, John von Neumann published what he thought was such a theorem, but it was quickly seen that the proof was based on a mathematical assumption that was not appropriate for the quantum world.

2.2.1 Models, Proofs, and Experiments

What we *can* do, and what most people view as sweeping and ambitious enough, is to create *models* that precisely describe a large class of theories about which we *can* provide impossibility proofs.

What Bell did (essentially) was to assume a model of what an Einsteinian theory should be, and then show that any theory consistent with that model would have mathematical consequences that *disagree* with certain predictions (and now real observations) of quantum mechanics.

2.2.1.1 Mermin's Model

We will discuss Bell's model and proof in a specific way that was introduced by David Mermin [68], in the context of proposed experiments that are generally called *EPR experiments*, first envisioned by Einstein, Podolsky, and Rosen [33] and later modified by David Bohm.

The generic experimental setup of an EPR experiment consists of a *central source* and two *detectors* on opposite sides of the source, placed at roughly the same distance from the source. The source will repeatedly generate and send out *two identical particles* (actually, either photons or atoms, but we don't need that detail) in opposite directions, and each particle will reach the detector on its side of the source (see Figure 2.1).

When a particle reaches a detector, the detector flashes a light that is either Red or Green. Further, any time the source emits two particles, each detector

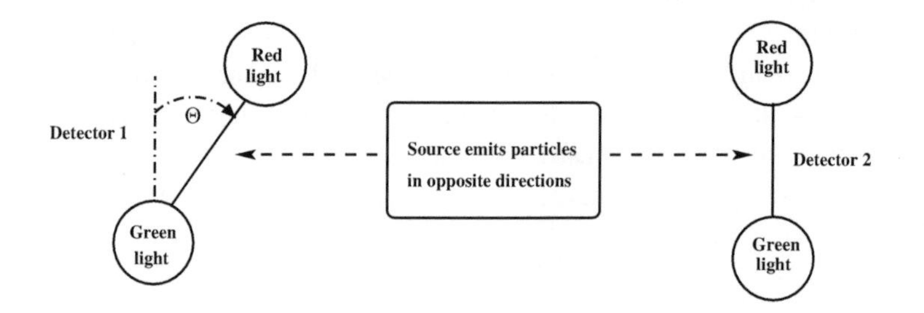

Figure 2.1 The generic setup for an EPR experiment. The source emits two identical particles in opposite directions. When a particle reaches a detector, either the Red light or the Green light will flash. We arbitrarily designate one detector as detector 1 and the other as detector 2. The experimental setup is shown from the perspective of an observer off to one side of the experiment. In that perspective, detector 1 is on the left and detector 2 is on the right. Detector 2 is shown oriented *vertically*, and detector 1 is shown rotated by an angle θ from the vertical, but the orientation of each detector is allowed to change each time a particle approaches it.

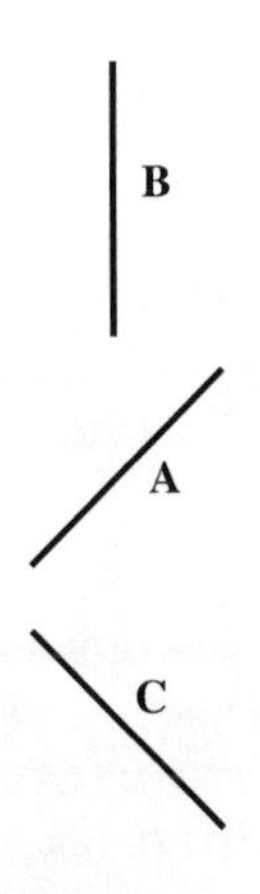

Figure 2.2 Each detector independently chooses one of three orientations, *A*, *B*, or *C*, just before a particle reaches it.

can be independently *oriented*, affecting the angle at which the incoming particle reaches the detector.[11] It will turn out that only *three* specific angles (call them *A*, *B*, *C*) are ever needed (see Figure 2.2).[12]

Some terminology We arbitrarily designate one of the detectors as detector 1 and the other as detector 2. We will use the phrase *joint orientation* when talking about the orientations of the *two* detectors. A joint orientation will be

[11] Actually, this is (essentially) the version of the EPR experiment used by Bell in 1964, not the original EPR experiment of 1935. In the original version of the EPR experiment, the detectors are always set vertically, which is sufficient for the point Einstein and collaborators were trying to make. They weren't trying to make the point that Bell made. He hadn't made it yet – he was only seven in 1935.

[12] The specific angles used in an EPR experiment are not needed in our discussion, but some readers may be interested. One set of angles that works is {0, 120, 240} degrees. That is, the orientations are equally spaced around a circle, so that the angle between any two orientations is 120 degrees.

Some readers might also be interested in the inner details of the detectors, that is, what the mechanism is that determines which color light flashes. However, explaining those details would involve hard-core quantum mechanics, and are not required for stating and proving Bell's theorem. For that task, we only need two statements, called *Q1* and *Q2*, about the *behavior* of the detectors, and not any details about their inner workings. Later, we will say a bit more about what in quantum mechanics leads to these two statements.

Although not needed for our focus on Bell's theorem, for readers who do want to learn a bit more about the physics of the detectors, I suggest the discussion of the detectors in chapters 2 through 6 of the lay-audience book by Daniel Styer [102]. But beware that his treatment of probability on page 34 is incorrect; it can be fixed by replacing the word "when" with the word "and" everywhere on that page. Also, be aware of the message developed in Exercise 6 in this chapter.

denoted by two letters, such as AC, CA, BB, and so forth, where the first letter specifies the orientation of detector 1, and the second letter specifies the orientation of detector 2.

When the source sends out a *single pair* of particles, this is called a *trial.* In a single trial, the detectors independently choose their orientations, and the particles reach the two detectors, which then turn on their respective lights. A series of trials is called an *experiment.*

Back to EPR Returning to the description of an EPR experiment: In each trial, each detector chooses a random orientation *with equal probability* from among the three choices of angles. That is, each detector randomly chooses an angle A, B, or C, each with probability *one-third.* This is like rolling a three-sided die.

Further, the orientation choices at the two detectors are completely uncoordinated: The choice at one detector has no influence[13] on the choice at the other detector. In more technical language, these choices are *independent.* In real EPR experiments, in order to further guarantee independence of the choices, a detector only chooses its orientation *shortly* before a particle reaches it, and the distances of the detectors from the source are large enough that the time between when one particle reaches its detector and when the other particle reaches its detector is too short to permit anything (even a message) to travel between the detectors (or the particles) at the speed of light.

Also, although it is already implied, for emphasis we state that the choices made in one trial are independent of the choices made in any other trial.

2.2.1.2 What Is Predicted, and Observed, in an EPR Experiment?

Quantum mechanics first predicted certain experimental results, and these have now been verified in many real experiments.[14] In the generic EPR experiment described above, the predictions and observations obey the following two statements:

Q1) In any trial where the two detectors have the *same* orientation (either both angle A, or both angle B, or both angle C), after the particles reach the detectors, the colors of the two lights will *agree* with each other. That color can sometimes be Red and sometimes be Green, but the colors will agree each time.

[13] At least, no influence of any type explained by classical (nonquantum) physics.

[14] The first experiment was done in 1972 [40] by John Clauser (then a postdoctoral student) and Stuart Freedman (a graduate student). In 2022, Clauser shared the Nobel prize in Physics for this work. Clauser and Freedman worked for several years to conduct the experiment, but by 2002, similar experiments could be done "in an afternoon" in an undergraduate teaching lab [31].

And,

Q2) In a "long" series of trials (i.e., a long experiment), the color of the two lights will *agree* with each other roughly *half* the time.[15]

Note that the two lights *might* agree with each other even when their respective detectors are oriented at *different* angles, but they *must* agree when they are oriented at the same angle. No further details of the experimental setup or results are needed. As an example, one experiment that agrees with statements Q1 and Q2 is:

$$AA, RR;\ \ BC, GR;\ \ AC, RR;\ \ AB, GG;\ \ CA, RG;\ \ BB, RR;\ \ BA, RG;\ \ BA, GR,$$

where each trial is represented by four letters. The first two letters (before the comma) specify the joint orientations of detectors 1 and 2, respectively; and the second two letters (after the comma) specify the colors of the two lights flashed by detectors 1 and 2, respectively. Each trial is separated by a semicolon. In this contrived example of eight trials, there are four trials where the colors of the lights agree (satisfying Statement Q2), and among those trials are all the trials where the detectors chose the same orientation (satisfying Statement Q1).

Bell's question restated in the context of EPR

> Is there an Einsteinian theory that can explain the quantum predictions (now experimental observations) summarized in statements Q1 and Q2?

Bell proved that the answer is essentially "no". But what does that even mean?

2.2.1.3 Modeling Einsteinian Theories

As stated earlier, to address such a question, we first need a *model* of an Einsteinian theory, that is, a way to precisely define what an Einsteinian theory can be, in the context of an EPR experiment. David Mermin in [68] proposed the following:

> After a particle has been generated and leaves the source, it behaves in the future **as if** it was given a precise rule telling it how to behave, depending on what it finds when it reaches a detector. The particle going in the opposite direction is similarly given a rule (not necessarily the same rule given to the first particle).

An example of such a rule given to a particle is:

> When you reach your detector, *if* it is oriented at angle A or C, *then* cause its light to flash Green; but *if* it is oriented at angle B, *then* cause it to flash Red.

[15] More technically, we can make the expected fraction of trials where the two lights agree, as close to one-half as we like, by extending the number of trials in an experiment.

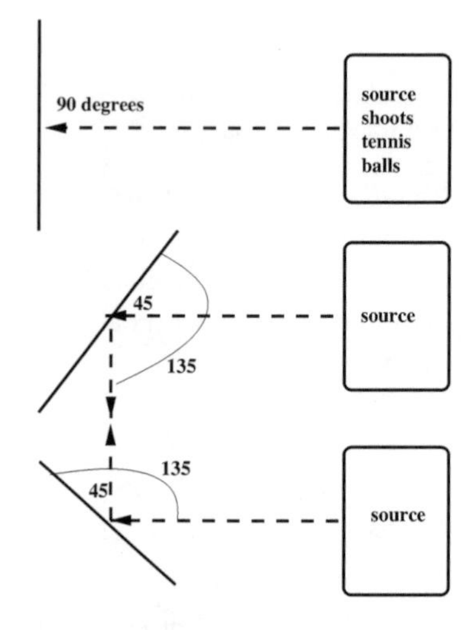

Figure 2.3 If the source shoots tennis balls instead of particles, the possible behaviors (bounces) of the balls, when they hit the backboard (detector), can be enumerated by the rule described in footnote 16.

Note that having such a rule does *not* mean that at the moment the particle leaves the source, we (or it) knows for sure what will happen when it reaches the detector. It only means that we can *enumerate* and *describe* the allowed future behaviors of the particle as a set of definite *if–then* type of instructions, as in the example above.[16]

It's just a model Of course, no one is suggesting that a particle is *literally* handed a 3-by-5 index card with a set of instructions (a rule) written on it every time it leaves the source. Instead, such rules are the way Mermin proposed to

[16] At first blush, this kind of *rule-encoded, case-based* model may seem awkward or contrived, but it really does capture the kind of typical *deterministic* behavior we expect in our normal (nonquantum) classical-physics world. For example, suppose that instead of particles, the source shoots out tennis balls; that a detector is now just a backboard that the balls bounce off of; that a backboard can be oriented at angles 90, 45, or 135 degrees, measured from the backboard to the incoming ball in the clockwise direction; and that instead of causing a light to flash, the observable phenomenon is the *angle* that the ball bounces off the backboard. See Figure 2.3. Then, we can capture the possible behaviors of the ball with the case-based rule: If you hit the backboard at an angle of 90 degrees, bounce off at an angle of 90 degrees; but if you hit at an angle of 45 degrees, bounce off at an angle of $180 - 45 = 135$ degrees; and if you hit at an angle of 135 degrees, bounce off at an angle of 45 degrees.

Admittedly, it is more natural to describe the possible behaviors of the tennis ball by referring to the *law of reflection for elastic collisions*: the angle that the ball leaves a surface is equal to the angle it approaches the surface. And, of course, that law is behind our stated rule. But, natural or not, the case-based rule correctly describes the three allowed behaviors of the ball in this system.

model what it means for a theory to obey Einstein's assumptions in the context of an EPR experiment. Still, let's call this the *index-card* model, to make it more visually memorable. Then, when we talk about an *index-card theory*, we mean a theory that obeys the index-card model.[17]

2.2.2 Now Impossibility Is Possible

With a precise model (the index-card model) in hand, we (well, actually Bell and Mermin) can finally *prove* something:

> *No* theory that obeys the index-card model can predict the behaviors (first predicted by quantum mechanics, and later observed) that are summarized in statements Q1 and Q2.

Bell changed "did not" to "can not" Before Bell's theorem, classical physics did not predict the behaviors in statements Q1 and Q2. But who was to say that it never could? John Stewart Bell – that's who.

Bell's theorem essentially changes "did not" to "can not". That is the essence of Bell's theorem, which we will now begin to prove.

Notation A rule will be specified by a string of three letters, where each letter is either R or G. This triplet specifies exactly the color light that the particle will cause the detector to flash, based on the orientation of the detector, that is, angle A, B, or C. For example, the rule in the above example is specified by the string GRG.
Since there are three possible angles for a detector, there are eight ($2^3 = 8$) such rules:

$$RRR,\ RRG,\ RGR,\ RGG,\ GGR,\ GRG,\ GRR,\ GGG.$$

But first, try to find index-card explanations Before we prove that there are no index-card theories that can make predictions Q1 and Q2 *together*, let's see if there are index-card theories that predict the two statements *separately*.

It is easy to think of index-card theories that predict Q1 by itself: Give both particles the rule RRR in each trial. Then, statement Q1 will be satisfied since

[17] I just made up the terms "index-card model" and "index-card theory", so if you use them elsewhere, don't expect anyone to know what you are talking about. The terms "instruction-set model" and "instruction-set theory" are commonly used, but I find them less visually compelling. And, as established terms, they already have certain established connotations. For clarity of my exposition, I want to start with a clean slate.

the two detectors will *always* flash Red. But, with this rule, statement Q2 will be violated.

Index-card theories that predict statement Q2 by itself, are also immediate: Give one particle the rule *RRR* in every trial, but give rules to the other particle that alternate between *RRR* and *GGG*, changing in each trial. Half the time, the particles will *both* be given rule *RRR*, and their lights will *agree* (in fact they will both be Red). But when one particle is given rule *RRR* and the other is given *GGG*, which happens half the time, their lights will *disagree*, so statement Q2 will be satisfied. Great, but do these rules satisfy statement Q1? No! Explain, or read on.

2.2.3 Bell's Theorem for the Index-card Model

Theorem 2.2.1 Every *theory that obeys the index-card model will either allow a violation of statement Q1 or predict a violation of statement Q2.*

Proof We first establish that in order to guarantee statement Q1 is *never* violated, the rule given to one particle *must* be given to the other particle in every trial, even though the index-card model does not explicitly require this.

To get the idea, suppose one particle is given the rule *RRR* and the other is given the rule *GGG*. If it then happens that the two detectors choose the *same* orientation, both angle *A* or both *B* or both *C*, statement Q1 will be violated.

In general, when the two rules given to the particles differ in *even one* position, say the first position (as in *RGG* and *GGG*), Statement Q1 will be violated if both detectors choose to orient at angle *A*.

Hence, to be sure that Statement Q1 is *never* violated in any trial, the rule given to one particle must be the same as the rule given to the other particle.[18]

A warm-up Now that we have established that in any trial, the particles must be given the same rules (or else Statement Q1 would not be guaranteed), let's consider a specific rule, say *RRG*, given to both particles.

We ask: For what joint orientations of the detectors will the two lights *agree*, when the two particles are given rule *RRG*?

[18] We have deduced this, but sometimes it is stated as an explicit part of the model to reflect the fact that the particles are *identical*.

We also note that in some EPR experiments the particles are not identical, but are perfect *opposites*, so the given rules are always opposites. For example, if one particle is given the rule *RGR*, the other particle will be given *GRG*. Conceptually and mathematically, this is a trivial difference, and we will always assume that the particles are identical in our analyses.

Certainly, the lights agree when both detectors are oriented to angle *A*, or both to angle *B*, or both to angle *C*. In the first two cases, the lights flash Red, and in the third case, the lights flash Green.

But, the lights will *also* agree if one detector is set to angle *A* and the other to angle *B*, that is, either joint orientations *AB* or *BA*. The lights will *disagree* in all other joint orientations of the detectors. So, when the rule is *RRG*, there are *five* joint orientations of the two detectors which result in the lights agreeing on the colors they flash; and there are *four* joint orientations where the lights will disagree.[19]

Hence, with the specific rule *RRG*, the two detectors will agree for *five-ninths* of the *joint orientations* that they can have. But, that is not really what we want to know. Instead, we want to know how *often* the two detectors agree. We need to know how often they agree in order to compare with statement Q2. Well, moving from *fractions of joint orientations* to *frequency of agreement* is easy.

Each of the nine joint orientations is chosen with roughly *equal* frequency in a long experiment, since each detector chooses its individual orientation with equal probability and is independent of the choice made by the other detector. So, regardless of the specific rule given to the particles, each of the nine possible joint orientations of the detectors occurs in roughly *one-ninth* $\left(\text{i.e.,} \frac{1}{3} \times \frac{1}{3}\right)$ of the *trials* in a long experiment. That is, each of the nine joint orientations occurs in (roughly) the *same* number of trials in a long experiment.

In those specific trials where rule *RRG* is given to the particles, the two detectors agree in five (out of nine) joint orientations, and since each joint orientation occurs in (roughly) the same number of trials, it follows that the detectors will agree in (roughly) five-ninths of those trials.

The general case The analysis just done for rule *RRG* was a warm-up, but the same argument used for *RRG* also works when any of the following index-card rules is selected:

RGR, *RGG*, *GGR*, *GRG*, *GRR*

The key point is that in each of these rules, *two* positions in the string have the *same* color, and one position has the *other* color. Hence, as in the argument for rule *RRG*, with any of these rules, there are five joint orientations of the two detectors where the lights will agree. And, since each joint orientation occurs equally often (roughly), when any of the rules *RGR*, *RGG*, *GGR*, *GRG*, *GRR*,

[19] There are a total of *nine* joint orientations of the two detectors: *AA*, *AB*, *AC*, *BA*, *BB*, *BC*, *CA*, *CB*, and *CC*.

or *RRG* is given to the particles, the two lights will agree in roughly *five-ninths* of those *trials*.[20]

The only index-card rules that we have not discussed are *RRR* and *GGG*. But, clearly, the two lights will always agree whenever either of these rules is given to the two particles.

So? We have shown that for *every one* of the eight index-card rules, in a long experiment, when we look at *those* trials where that *specific* rule is given to the two particles, the detectors' lights will agree in (roughly) *five-ninths or more* of *those* trials. So, in any long experiment, no matter how often each of the eight rules is given to the two particles, the lights will agree (roughly) five-ninths (or more) of the *time*.[21]

The punch line For *any* theory that obeys the index-card model, if no violation of statement Q1 is allowed, the theory *must* predict that the lights agree (roughly) *five-ninths* or more of the time, in a long experiment. But in the predictions of quantum mechanics (as summarized in statements Q1 and Q2), the lights will *only* agree (roughly) *one-half* of the time, in a long experiment.[22] This implies that

> *No* theory that obeys the index-card model can make *all* the same *predictions* that quantum mechanics makes. Such a theory is *impossible*.

■

We did it! We have proven a Bell *impossibility* theorem.

Another conclusion Above, we established that the *predictions* of any index-card theory must either violate Statement Q1 or violate Statement Q2. But, we can view that result in a different way. *If* Statements Q1 and Q2 correctly characterize quantum phenomena in an EPR experiment, then

[20] If this is not clear, just *separately* rerun the analysis we did for rule *RRG*, for each of the five rules above.

[21] Notice that, we have said nothing about how often each of the eight rules is given to the particles. Different index-card theories could propose different ways that specified rules are assigned. This might seem at first to be a problem for the proof, but it isn't since the five-ninths (or more) result holds for *every* one of the eight index-card rules. It doesn't matter how often each specific rule is assigned.

[22] You may ask: Is five-ninths really that different from one-half? Yes! In a long series of trials, the difference between five-ninths and one-half is very *significant*, that is, much much larger than one would expect due to normal statistical fluctuations in the random choices of the detectors' orientations.

No theory that obeys the index-card model can explain *all* quantum *phenomena*. Such a theory is *impossible*.

2.2.4 Where Is the Inequality?

Bell theorems are often referred to as Bell *inequalities*. In Chapter 3, we will explicitly discuss a general Bell inequality, but one can extract an inequality from the proof of the Bell theorem we have discussed here:

In the index-card model of an EPR experiment, if in each trial, both particles are given the same rules (which can vary between trials), then in a "long" experiment, the proportion of trials where the detectors lights agree in color should be greater or equal to $\frac{5}{9}$ (about 55.5%).

Note that this is a statement about the *statistics* that we should see in a long EPR experiment, if classical physics and the index-card model apply.

Bell's inequalities always tell us something about what to expect in the normal, classical, nonquantum world.

This is nicely stated in [79]:

Bell's inequality is **not** a result *about* quantum mechanics ...

Instead, it results from analyzing

... our common sense notions of how the world works – the sort of notions Einstein and his collaborators thought Nature ought to obey.

Bell's inequality is not about quantum mechanics per se, but its *relevance* is. Bell's inequality highlights the different *statistical* results that would be expected under the index-card model, compared to the expected results under quantum mechanics. So, when the results of an experiment *violate* the inequality, we conclude that some nonclassical phenomena have occurred, and that the quantum world does not conform to "common sense notions of how the world works".

2.3 Why Is This Shocking and Profound?

We just proved an *impossibility* theorem, a Bell theorem, using the index-card model. If you only want to see how impossibility is modeled and proved, you can stop here. But the story is much deeper, and we want to understand why Bell's theorem is *shocking* and *profound*. What makes it a candidate for "the most profound discovery of science"?

First, we asserted that the index-card model captures what we called "Einsteinian" theories, and if that is true, then the Bell theorem we proved actually says that *no* Einsteinian theory can explain all quantum phenomena. More generally, the index-card model is implicit in classical physics, implying that no theory based on classical physics can explain all quantum phenomena. We will amplify this view below.

Second, Bell's inequalities show *how* experiments can address the question of whether a truly bizarre, and technologically important, quantum phenomenon called *entanglement* really exists or not.

So, we will next explain more completely what Einsteinian theories are and argue that the index-card model *does* indeed capture Einsteinian theories, and reflects classical physics. And, we will discuss entanglement and how Bell's theorems relate to it. We will do this by discussing more of the history of quantum mechanics and Einstein's objections to it. Then, in Chapter 3, we will more explicitly discuss Bell *inequalities* and how they are used in experiments to support the conclusion that entanglement is real.

2.3.1 The Backstory

The background to Bell's theorem (published in 1964) was the (then) unresolved great debate (or argument) that Einstein had with Niels Bohr (the greatest champion of orthodox quantum mechanics), over several years, about the nature of reality, physics, theory, the universe, and, particularly, the correctness and completeness of quantum mechanics.

Quantum theory actually builds on earlier work of Einstein (in 1905), but Einstein had great reservations about the formal, mathematical theory, that is, quantum mechanics, that was articulated in the 1920s. Historians of this debate are a bit unsure about what exactly troubled Einstein at the start (and are extremely unsure about what Bohr was saying[23]), but by 1935, Einstein was able to more fully articulate his objections to quantum mechanics. In that year, Einstein, with two younger associates, published a paper [33] that came as a "bolt out of the blue", forcing Bohr to drop everything he was doing and focus on a reply (which few people claim to understand).

In the 1935 paper, Einstein, Podolsky, and Rosen (EPR) showed that *if* one applies the quantum mechanical theory that Bohr was championing to an experimental setup that they devised (now called an *EPR experiment*), *then*

[23] "Writing was a painful process for Bohr, and nearly impossible for him to accomplish without help ... Bohr's writing was tortuous and obscure, with famously long and convoluted sentences." [11]

there would be a truly astonishing physical phenomenon that had never even been imagined before.[24]

Entanglement That phenomenon was later called *entanglement*, which some physicists now say is **The central characteristic** of the quantum world. And, as we will discuss more later, it is entanglement that leads to the quantum mechanical predictions summarized in Statements Q1 and Q2.

Entanglement means that under certain conditions, two particles can physically interact, becoming "entangled",[25] and then no matter how *distant* they are from each other, or how much time has passed since their entanglement, some of their quantum properties will be *linked*. In some types of experiments, quantum properties will be *perfectly correlated* (i.e., identical), and in other types of experiments, they will be perfectly *anti-correlated* (i.e., opposite).

The quantum properties of concern are a bit esoteric and unfamiliar, and the way objects become entangled can be complex. A real discussion of entanglement involves hard-core quantum mechanics, which is way outside the scope of this book and is not needed for our purposes. Instead, we will illustrate the concept of entanglement with a *fictional* analogy.

2.3.2 A Fictional Example of Entanglement

This fictional example is an *analogy*, in order to help describe the implications of entanglement.

Suppose two Jefferson nickels touch as they roll off the assembly line at the Philadelphia mint and (somehow) became entangled. Then, the two nickels drop into two separate boxes, each large enough to allow a coin to flip over when shaken. One box is flown to Mars (a rough ride) while the other stays on Earth. When the coins dropped into the two boxes, nobody looked to see whether a coin landed heads up or heads down in a box. However, it is known from observing other coins, that about half of the nickels coming out of the mint are oriented with heads up and about half with heads down, and the distribution of coins with heads up or head down looks totally random.

When the box sent to Mars arrives, the two boxes are opened at roughly the same time. Assuming that the "which-side-is-up" property of a coin is a

[24] Hey, Einstein was a pretty smart guy – they didn't call him "Einstein" for nothing.

[25] And, as part of his work leading to his share of the Nobel prize, Anton Zeilinger and coworkers showed how particles can become entangled even when they never physically interact.

quantum property, here is where people (and Martians) encounter the implications of entanglement:

> When the two nickels in the boxes are examined, they will *both* be observed to be heads up, or *both* observed to be heads down. This perfect correlation is despite the rough ride to Mars, the time that has passed since the coins became entangled, the distance that separates them, and the fact that if coins coming out of the mint are examined, the face that is up is essentially *random*.

Of course, agreement on one pair of coins doesn't seem so surprising, but suppose that instead of just one pair of entangled nickels, a *billion* pairs of entangled nickels are created (how much is that in real money?). One nickel in each pair stays on Earth, and one goes to Mars. When the two nickels in the first entangled pair are examined (at roughly the same time), the results are perfectly correlated. Then, two more entangled nickels are examined (at roughly the same time), and again the results are perfectly correlated. Continuing in this way, in each of the billion pairs of entangled coins, the results are *always* perfectly correlated – either both coins are observed heads up, or both are observed heads down. That is entanglement, and it is strange.

Meanwhile, back on planet Earth (or Mars) Filling in a bit more detail, in roughly half the entangled pairs of nickels, the two coins are observed heads up, and in roughly half, the two coins are observed heads down. And, which pairs are heads up and which are heads down looks completely random, consistent with what is observed of coins coming out of the mint.

So, a person looking at just the coins from the entangled pairs on Earth, or just the coins on Mars, would not see anything strange, compared to coins that are not in an entangled pair. It is only when people *compare* the results for the billion entangled pairs of coins that they see the strange perfect correlations that are the hallmark of entanglement.

And if entanglement is real, it is "spooky" One last, very important point. We said that the two coins in any entangled pair are examined at roughly the same time, and far from each other. Why was that so important?

It is a *central tenet* of quantum mechanics that a quantum property of an object (a nickel in this case) *does not exist* before it is observed. So, which side of the coin is up is only determined (in some unexplained random way) at the moment that the coin is observed. Hence, the two entangled coins could not use the time before the observations to *collude* on which side will be up. And since the coins are observed at roughly the same time, any difference in the times that the two coins are examined will be smaller than the time it takes light to travel between Earth and Mars. So, if one coin is examined slightly before the other one, there isn't enough time for "reasonable" communication (i.e., no faster than the speed of light) to convey the result of the *first* examination to the other coin, *before* it is examined.

But, the entangled coins are always observed to be perfectly correlated. So, based on our normal, classical-physics intuition, it seems that there must have been some kind of *abnormal, faster-than-light* communication between the coins that enforces the perfect correlation: Entanglement suggests instantaneous *communication* or *influence*.[26] Einstein described this as "spooky".[27]

2.3.2.1 Recapping, and Continuing

Recapping, the 1935 EPR paper showed that quantum mechanics predicts entanglement, so *if* quantum mechanics is correct, *then* entanglement is a real phenomenon. And, entanglement seems to imply *instantaneous* communication or influence across any distance (It seems spooky!).

But, in 1935, and in fact until 1964 (when Bell published his paper), no one had any idea how to determine if entanglement was real or not. Classical physics did not predict or explain entanglement, but who was to say that it couldn't ever do so? Maybe some future "Einstein" would find a way to explain those behaviors using only principles of classical physics.

The real Einstein didn't buy it Of course, Einstein did not believe entanglement was real. Quoting from the original EPR paper:

> No reasonable definition of reality could be expected to permit this. [33]

Einstein presented the *if–then* argument as a way to convince others that there was something incomplete about quantum theory. Although he didn't use the word "entanglement", he expected that people would say something like: "Existing quantum mechanical theory implies entanglement, and entanglement is too unreasonable and spooky to be true, so there must be something incomplete (or maybe incorrect) about the current quantum mechanics."

But historically that didn't happen. Instead, over time, quantum physicists said "Wow, great, entanglement is real and quantum mechanics is even more amazing than we knew (thanks be to Einstein for discovering it). Now, let's go create entanglement and exploit it!" – and they are doing that today.

As summarized in [79]:

[26] But, we will see that this is not actually the orthodox quantum mechanical *explanation* for the correlation.

[27] *Spukhafte Fernwirkungen*, which sounds pretty spooky to me.

> We're stuck with the counter-intuitive nature of quantum mechanics. Of course, the proper reaction to this is glee, not sorrow!
>
> Bell's inequality [theorem] teaches us that entanglement is a fundamentally new resource ... that goes essentially *beyond* classical resources; iron to the classical world's bronze age. (italics are original)

2.3.3 A Bit More about EPR Experiments

Now that we have introduced entanglement, we can make explicit a key detail of the EPR story.

> When the source creates the two particles it sends out in an EPR trial, those particles are (predicted to be, by quantum mechanics) *entangled*.

And, it is entanglement that leads to the prediction in Statement Q1: When the entangled particles encounter the same environment (e.g., orientation of their respective detectors), they behave identically, causing the same color light to flash.

Further, quantum mechanics predicts that when the entangled particles encounter different environments (different detector orientations), their behaviors will not always be identical, but the extent of their correlation will depend on how different the environments are. That principle, and the specific angles used in the EPR experiment, leads to Statement Q2. This is explained a bit more in Exercise 4.

So, having proved that it is impossible for any theory that obeys the index-card model to predict Statements Q1 and Q2, and the fact that in quantum mechanics, entanglement leads to predictions Q1 and Q2, we can also conclude that no index-card theory can predict entanglement. Further, now that many real experiments have been done, Statements Q1 and Q2 are not just predictions – they are summaries of consistent observations. Bell's theorem then implies that there are physical phenomena that fundamentally fall outside of the explanatory power of classical physics.

2.3.4 Einstein and the Unstated Assumptions

Einstein showed that quantum mechanics implies things he found "spooky." But, why was Einstein so resistant to quantum theory?

Einstein insisted that a more intuitive theory (more like classical physics) *should* be possible to explain the quantum world and the unintuitive quantum behaviors that had already been observed by 1935. He insisted that any theory about quantum phenomena should make the same assumptions that are implicitly made in classical physics. There are two of them.

2.3.4.1 First Assumption: Physical Reality

Einstein believed that there is a "physical reality" where physical objects and phenomena exist "out there", whether anyone observes them or not. Further, the goal of physics is to discover and explain that physical reality.

Quantum mechanics takes the opposite view: that there is no "out there" and that properties of physical objects are only *created* when they are *observed*. Bohr is reputed (see [73]) to have said:

> It is wrong to think that the task of physics is to find out how Nature *is*. Physics concerns what we can *say* about Nature.

Einstein's insistence that physical reality exists, whether it is observed or not, is one of the (self-evident, unstated) assumptions in classical physics, and one that is accepted without thought by most of humanity (other than physicists and philosophers). When you look at a car and see that it is black, you generally have no doubt that it was black a fraction of a second before you looked at it. Quantum mechanics disputes that.

Orthodox quantum mechanics says that a fraction of second before the car was observed, *it had no color*. It's not just that we couldn't know the color of the car before we looked, it's that the car really had no color before we looked. It may even be that the car didn't exist.

A revealing story There is a story in which Einstein was walking with another physicist, Abraham Pais, discussing the meaning of objective reality, when Einstein turned to him and asked "Do you really think the moon is not there when no one looks at it?"

We don't know what Pais replied, but the orthodox view of quantum mechanics is clear: "... physical properties have in general no objective reality independent of the act of observation" [68]. Nothing is "there" until it is "observed" (although not necessarily by anything alive). That was, and is, the orthodox view, but in Einstein's time, they had no idea *how* that view could be proved or disproved.

2.3.4.2 Second Assumption: Locality

The second intuitive assumption from classical physics that Einstein insisted on, is that when two objects are physically separated and something affects the first object, that effect *cannot instantaneously* be felt by the second object – it takes time for the "effect", or "message", or "influence", or "force" to travel from the first object to the second. Moreover, the more distant the two objects are, the longer it takes for the effect to travel between them. More precisely, Einstein insisted that an effect cannot travel from one object to another faster than the speed of light.

Certainly, objects at the second location can be affected by what happens at the first (think of a bullet fired at the first location hitting an object at the second location), but the effect is not instantaneous. Today, Einstein's position is more precisely stated as: *information* about the event at the first object cannot be transmitted to the second object faster than the speed of light.

But quantum mechanics seems to imply that when two objects are entangled, an effect felt on one object *will be instantly* felt by the second object, even if those two objects are at opposite sides of the universe.[28]

2.3.5 What Kind of Theory Did Einstein Insist On?

Given the above discussion, we see that the kind of theories that Einstein insisted on must assume that objects have *fixed, real* properties, even before anyone has observed them; and that what happens to an object at one location in space cannot instantly influence what happens to an object at a different location in space.

Although Einstein did not use these terms, his view is now called *local realism*. The word "realism" in this context means that physical properties and

[28] Actually, this is an attempt to translate quantum mechanical theory (which can only be correctly stated mathematically) into everyday language. But some physicists do use such language. From [112]:

> Entanglement describes the phenomenon that two (or more) particles (or systems) may be so intimately connected to each other that the measurement of one *instantly* changes the quantum state of the other, no matter how far away it may be. (italics added)

However, what physicists more often say, when trying to translate from mathematics to everyday language, is something even more unnatural (to me). They say that quantum theory *does not* imply that when two objects are entangled, an effect on one is instantly felt by the second. Instead, they say that what we think of as *two* entangled objects are actually connected parts of the *same* object, even if the two parts are very distant from each other, and even if there is no part of the object in between. And because the two parts are part of the same object, nothing needs to go *between them*. Something that happens at one part of the object will instantly affect other parts of the object. Quoting from [18]: "... even when you place those particles on opposite sides of the galaxy, they will behave as if they are actually one particle."

To me, this seems like an unconvincing semantic dodge to avoid saying that an effect can be *instantly* felt, no matter how far apart the *two* objects are. But, that is part of what makes the mathematical theory of quantum mechanics so unintuitive and difficult to translate into normal language. In my opinion, the mathematical definition of entanglement (which is way outside of the scope of this book) is actually easier to grasp and accept than the attempted translations of it into natural language.

objects exist whether anyone observes them or not. The word "local" means that an effect on one object cannot be felt instantly by a distant object.

Einstein insisted that there must be a (yet undiscovered) theory (i.e., an explanation) of physical behavior in the quantum world that obeys the realism and locality assumptions. That would be a *local realistic* theory of quantum behavior. Quoting from [32]:

> The three premises[29] which are often assumed to have the status of well-established truths, or even self-evident truths, form the basis of what I shall call local realistic theories of nature.

Given this, we see that Bell's larger question was:

> Can a local realistic theory explain all predictions made by quantum mechanics? Or is such a theory *impossible*?

Bell's mathematical theorem convinced most physicists that it is impossible.

2.3.6 But How?

How does the (purely mathematical) Bell theorem that we proved in Section 2.2.3 (concerning precise index-card models) imply the more abstract claim that no local realistic theory can explain all of quantum mechanics?

Entanglement and Statements Q1 and Q2 As stated earlier, entanglement leads to the derivation (in quantum mechanics) of Statements Q1 and Q2. So, proving that no index-card theory can predict both Q1 and Q2 means that no index-card theory can predict entanglement. And, the index-card model is implicit in classical physics. Further, most physicists (and philosophers of physics) accept that models (essentially) like the index-card model capture what it means to be a local realistic theory.

Index-card theories are local realistic theories Certainly, theories that obey the index-card model do obey local realism. In the index-card model, a particle is given a *definite* rule when it leaves the source, so that its *possible* future behaviors are fully characterized before anyone *observes* its behavior.[30] That is the essence of Einstein's *realism*.

[29] These are locality, realism, and a third premise that is more philosophical than physical called *inductive inference*, which means that "legitimate conclusions can be drawn from consistent observations" [32].

[30] Remember that the actual behavior of the particle depends on what the particle finds at its detector. We can say that its behavior is to feel the angle of the detector and then to cause the detector to flash the light specified by the rule the particle was given.

And, by following the rule given to it, the behavior of one particle when it reaches its detector is uninfluenced by the behavior of the other particle when that particle reaches its detector at roughly the same time. That is the essence of Einstein's *locality*.

Hidden variables Accepting the other direction, that any local realistic theory fits into the index-card model, is a more subjective matter. But, physicists generally accept the view that local realism implies that physical objects (e.g., particles) must have (unknown, or hidden) properties that guide the behavior of the objects. They use the catch-all term *hidden variables*[31] for the hypothetical properties that the objects might carry to instruct them on how to behave in different situations. The hidden variables hypothesis is that

> ... each particle carries some property or instruction that determines the specific measurement result [e.g., color of light], one instruction for each possible measurement setting [e.g., detector angle]. [112]

> ... We call these additional properties each particle carries *hidden variables* ... all they do is determine the measurement result ...
> ... the assumption of such properties or instructions is rather reasonable. [112]

The assertion that objects have hidden variables is essentially the index-card model, supporting the assertion that any local realistic theory can also be viewed as an index-card theory.

Recapping The Bell theorem that we (rigorously) proved is a precise mathematical statement based on a precise model that we called the *index-card* model. Physical theories that obey the index-card model are local realistic theories. The converse claim, that any local realistic theory obeys the index-card model, is essentially *equivalent* to the accepted view that local realism implies the existence of hidden variables.

> Local realism is implied by the existence of *local hidden variables* ... the reverse implication is also true: local realism implies that we can construct a local hidden variable (LHV) model for the phenomenon under study. [43]

[31] Which is a term I dislike and find misleading. But I didn't make it up. I thought of not mentioning the term, but it is so widely used in almost all discussions of EPR and Bell that in the end I had to introduce it. But now that I have, I will try not to use it much.
John von Neumann used the term "verborgenen Parametern" which translates to "hidden parameters" in his 1932 book on quantum mechanics: *Mathematische Grundlagen der Quantenmechanik*. That term makes more sense to me than "hidden variables".

So, we conclude that the index-card model (which allows us to state and prove precise impossibility) is a valid, concrete way to model hidden variables and the consequences of local realism.

2.3.7 Now We Can See What Is so Shocking and Profound

Accepting the view that any local realistic theory fits into the index-card model means that the Bell impossibility theorem we proved implies that

> *No* local realistic (classical-physics) theory can make all the same predictions that quantum mechanics makes, and if quantum mechanics correctly predicts quantum phenomena (such as entanglement) no local realistic theory will *ever* be adequate to explain the quantum world.

In particular, quantum phenomena violate either *realism* or *locality*, or both. Violating realism means that properties of an object *do not exist* until observed; and violating locality means that "... the act of measuring one particle could [can] *instantly* influence the other one" [112].

> Bell showed that it is not possible to understand the phenomenon of entangled systems if one starts from rather "reasonable" assumptions [local realism] of how the world should work, assumptions that one might even be tempted to call self-evident. [112]

And quoting from Roland Hanson [18], who conducted one of the most recent and definitive EPR experiments:

> If you start with Einstein's concepts of locality (local forces) and realism (real properties) you end up at Bell's inequality. And Bell's inequality was violated in our experiment. So we have to drop either locality or realism ...

Similarly,

> The assumptions that went into the derivation of the Bell inequality [theorem] are the assumptions of local realism. So, the conclusion is that the philosophical position of local realism is *untenable*. [112]

That is shocking! Bell's theorem together with many experimental results shows that the universe behaves in ways that defy our (and Einstein's) "self-evident" notions of realism, or of locality, or both. More forcefully, Bell's theorem implies that "reasonable" theories based on local realism can *never* predict or explain all quantum phenomena.[32] Bell's theorem dashed Einstein's

[32] We should note that there is some disagreement about the necessity of the implicit assumption of realism in the physical underpinnings of Bell's theorem [11]. One view is that locality itself implies realism, so there is no need to make an assumption of realism. Moreover, there is

hopes (although he was dead by then) that such theories would someday be found. As John Bell himself put it:

> ... it is a pity that Einstein's idea doesn't work. The reasonable thing just doesn't work. [15]

Similarly:

> Einstein said that if quantum mechanics were correct then the world would be crazy. Einstein was right – the world is crazy. Daniel Greenberger

> Nature isn't classical, dammit. Richard Feynman

Bell's theorem ensured that

> ... logic that would be considered a psychiatric disorder in other fields, is the accepted norm in quantum theory. [54]

That and how Quoting from [104]:

> Until 1964 it was assumed that one could always construct a [local] hidden variable theory that would give all the same results as quantum mechanics. In that year, however, John S. Bell pointed out that alternative theories based on Einstein's locality principle actually yield a *testable* inequality that differs from the predictions of quantum mechanics. (italics added)

Before Bell's theorem and inequalities, it was not even imagined how such a conclusion could be established, and because of that, most physicists took little interest in the disagreement between Einstein and Bohr. Without a testable experiment, the "great debate" just looked like a philosophical debate with no tangible consequences.[33] But,

some disagreement about the very definition of "realism" [105]. However, if Bell's theorem does not require the implicit assumption of realism, but only locality, then its implications are even stronger than what we have discussed here. So, however this disagreement turns out, it will not call into question the discussion of Bell's theorem in this chapter. It only means that there might be an even stronger interpretation of what Bell proved.

[33] The original EPR paper had little initial impact - Google Scholar shows that it was cited in subsequent publications a total of only 15 times in the 10 years after its publication, and only a total of 69 times from 1935 until Bell's paper in 1964 (a span of nearly 30 years). The publication of Bell's paper didn't immediately impact the visibility of the EPR paper, and even in 1984 (20 years after Bell's paper, and nearly 50 years after EPR) it was cited only 59 times that year. The first year it was cited more than 100 times was 1992, after which its citation count began to steadily climb. In every year from 2012 to 2022, it was cited close to, or more than, 1,000 times. It is now Einstein's most cited paper, with close to 23,000 total citations – more in 2021 than in any prior year, with a similar count in 2022. Bell's EPR paper was also cited more than 1,000 times in 2022, but was ignored for many years before that.

The physicist, and friend of Einstein, Abraham Pais wrote the most authoritative and detailed scientific review [82] of every paper that Einstein (co)authored (about 300 of them). But in that 575-page book, the EPR paper is discussed in less than one page. Pais (in 1982!)

> In 1964, with the stroke of a pen, the Northern Irish physicist John Stewart Bell demoted locality from a cherished principle to a testable hypothesis ... [19]

Bell's impossibility theorem shows that physical theories embodying our self-evident truths are incompatible with quantum phenomena. That is why some say that Bell's theorem is the "most profound discovery of science."

With only a little arithmetic and logic, in his spare time And, Bell did this work quietly in his spare time – he was not employed as a theoretical physicist – he was more of an engineer.[34] The following misquote is from John Bell:

> During the week I was a quantum mechanic, but on Sundays I had principles.[35]

The story continues in the next chapter In this chapter, we explicitly proved an impossibility result: *a Bell theorem*, which is also referred to as a *Bell inequality*. But its connection to a Bell inequality was a little indirect.

In Chapter 3, we will explicitly prove a general Bell *inequality*, which allows experimentalists to recognize when quantum phenomena violate realism or locality and to recognize that entanglement is likely involved. That inequality will also provide another impossibility proof and will help to further explain why nonrealism or nonlocality cause Bell's inequalities to be violated.

And, similar to what we saw in this chapter, only arithmetic and simple logic will be needed in those proofs.

basically says that the paper only shows Einstein's distrust of quantum mechanics, and that the paper had no content of lasting importance. Quoting from [82] (page 456):

> The only part of this article that will ultimately survive, I believe, is this ... phrase ['No reasonable definition of reality could be expected to permit this'], which so poignantly summarizes Einstein's views on quantum mechanics in his later years. The content of this paper ... simply concludes that objective reality is incompatible with the assumption that quantum mechanics is complete. This conclusion has not affected subsequent developments in physics, and it is doubtful that it ever will.

And it is interesting that Pais (and others) held this view even 18 years after the publication of Bell's paper, and 10 years after the first actual EPR experiments were done. Pais and others did not yet see the importance (both conceptual and practical) of entanglement. This is a spectacular example of how difficult it is to predict the future value of knowledge. I think Pais would be very surprised to see how different the future actually turned out, as the EPR paper definitely did affect "subsequent developments in physics".

[34] And being somewhat outside of theoretical physics, his two first papers on the topic were largely ignored. One was published in an obscure journal that folded after a year, and the other sat on an editor's desk for two years until better known people published the same result.

[35] Well, actually he didn't say quantum "mechanic" – He said quantum "engineer". But how can one pass up such a lovely pun – so I had to change the quote.

2.4 Exercises

1. Early in this chapter we asked whether impossibility is possible. Foreshadowing the later chapters that relate to formal logic, let's have a little fun now. Consider the following argument: If impossibility is not possible, then it is impossible; so, something is impossible, hence impossibility is possible. In short, if impossibility is *not* possible, then impossibility *is* possible (in fact, necessary). "That's logic" said Tweedle Dum and Tweedle Dee.

What do you think of this argument? Does it make your head hurt?

2a. In the description of an EPR experiment, we said that the angle of a detector was chosen shortly before the particle reaches it. Why is this so important? How would the logic of the experiment, and the conclusions drawn from it, change if the angles of the two detectors were chosen before the particles were shot toward the detectors?

2b. Also, the angle of a detector is chosen *randomly* each time a particle approaches. Why is it important that the choice be random? How would the logic of the experiment and the conclusions drawn from it change if the choices were not random?

3. Orthodox quantum mechanics holds that a property of an object being measured (observed) does not exist until it is observed. How does that assumption affect the logic of an EPR experiment and the conclusions drawn from it?

4. Quantum Statement Q2 says that in a long EPR experiment, the color of the two lights will agree roughly *half* the time.

That statement is actually a consequence of the more precise quantum-mechanical statement:

Q2′) In a long EPR experiment, if we focus on those trials where two detectors have *differing* orientations (say, one is set at angle A, and the other is set at angle B), after the particles reach their detectors, the colors of their two lights will agree with each other in roughly *one-quarter* of those trials.

Another way of saying this is that in any trial where two detectors have different orientations, the colors of their resulting lights will agree with a probability of one-quarter.

Explain how Statement Q2′ implies Statement Q2. Hint: It helps to consider how often the detectors' orientations agree (when Statement Q1 can be applied) and how often the detectors' orientations differ.

5. The index-card model of an Einsteinain theory is a *deterministic* model. For example, one rule states: "if you find the detector at angle A, cause it to flash a

Red light." That is a deterministic rule – saying *exactly* what should happen *if* the particle finds the detector at angle A.

In contrast, an example of a *nondeterministic* rule for a particle is: "If you find the detector at angle A, roll a six-sided die, and if number 1 or 2 comes up, cause the detector to flash its Red light; but if any number 3 through 6 comes up, cause the detector to flash its Green light."

Rules like this can still satisfy local realism: A particle leaving the source has complete instructions *at that moment*, on how to act when it reaches the detector. And, locality can again be assumed because of the distance between the detectors and the fact that they choose their orientations only shortly before a particle reaches it.

But the rules are no longer deterministic because they involve an element of chance (rolling a die). So, this is a model of local realism that is less constrained than the one we discussed.

Can such a nondeterministic model explain Statements Q1 and Q2 together, or can we again establish that such a local realistic theory is impossible, even if it is not deterministic? Explain fully.

6. It is surprising that some expositions of Mermin's proof of Bell's theorem omit a key element of the proof. (For the sake of civility, I will not name names.)

We generally agree that the probability that a fair coin comes up heads in a coin flip is one-half. Is the reason for that because there are only two possibilities – either it will or it won't come up heads?

Is the following reasoning correct? The probability that I will win the next million-dollar lottery is one-half, because there are only two possibilities – either I will or I won't.

Back to the proof of Bell's theorem. A mistake I have seen several times is to not use, in the proof, the fact that in each trial of an EPR experiment, the orientation of each detector is chosen with equal probability (i.e., one-third) and independent of the other detector. The random choice of orientation is mentioned in the description of the EPR experiment, but that fact is not used in the proof of the theorem.

Essentially, those "proofs" assert that if a specific result (say identical colored lights) occurs in at least two-thirds (say) of the possible *joint orientations* of the detectors, then the probability that the specific result will occur in a trial is at least two-thirds. It is, but a correct, complete proof has to explicitly give a reason for it. Relate this to the questions about coin flips and the lottery.

Explain why the proof does not work if one makes the above mistake. Hint: What happens to the analysis if an index-card theory uses the rule RRG in each trial (which is a permissible, although silly, index-card theory), but the joint orientation of the detectors is AC in every trial (which is permissible unless the proof includes some reason that makes it impermissible)?

Find exactly where in the proof of Bell's theorem presented in this chapter the above issue was correctly handled.[36]

7. **Discussion question** Is it easier for people to walk forward than to walk backwards? Who taught you that fact? Did your parents or teachers lecture you about it? Have you ever heard that fact explicitly stated? What point am I making here, and how is it related to the topic of this chapter?

8. **Discussion question** Suppose there is a world (Flatland) where people can only move and rotate in two dimensions, on the plane. They cannot flip over using the third dimension that points in/out of the plane, and they cannot ever move out of the plane.

People in that world know that to walk between two points, A and B, they must traverse an unbroken path on the plane from A to B. If they mark that path as they walk, then there will be an uninterrupted curve drawn in their world, from A to B.

Now, the people in Flatland actually live in three dimensions, as we do, but they only *perceive*, and move in the two dimensions on the plane, and they don't know about the third dimension. But suppose one day some outside hand (living in three dimensions) grabs a person at point A and moves them in the third dimension to point B. People in Flatland will perceive the moved person as disappearing from their world at point A, and popping back into their world at point B, without traversing any path from A to B. What a shock!

Since the people in this two-dimensional world cannot perceive the third (in–out) dimension, how would they explain the impossible movement? Could they even imagine and articulate what a third dimension is and how it relates to the dimensions they perceive? Maybe those people would be smart enough to develop a mathematical theory that allows such movements, but would it ever really "make sense" and feel convincing?

The point of this exercise is to really try to imagine yourself as one of those two-dimensional people, and feel how confused you would be if you saw a person popping out of, and into, your world.

[36] OK, some people may find my view on proofs overly pedantic, but one of my responsibilities, when I was employed, was helping students understand what a real proof is. One thing I repeated is that if the description of a system includes a fact that is not explicitly used in the analysis of the system, then either the analysis should remain correct if the statement of the fact is *reversed* or the analysis is incomplete and should be fixed. Which is it in this case?

But, what does this thought experiment have to do with us? We live in all three dimensions, and we know we do. Surely, we would never be confused the way Flatlanders are.

Do you think that popping out of and into Flatland is more disturbing to Flatlanders than entanglement is to us?[37]

[37] If you enjoyed these questions, and have never read the book *Flatland* by Edward Abbey, I recommend reading the second part of the book. The first part of the book describes Flatland people and society, and written in the 1800s, it is very sexist. But literary historians believe Abbey was using fiction to ridicule sexism (and class structure). The second part of the book concerns the difficulty that two-dimensional people (who actually live in three dimensions) have in understanding their physical world.

3
Enjoying Bell Magic: With Inequalities and Without

The EPR experiment is as close to magic as any physical phenomenon I know of, and magic should be enjoyed. David Mermin [68]

In Chapter 2, we proved a Bell impossibility theorem and extracted an inequality from its proof. However, that inequality was stated only in the context of a specific EPR experiment. In this chapter, we first discuss an explicit Bell Inequality that has *universal application*, or at least, application in the "reasonable" parts of the universe. (In fact, it is when the inequality is violated that we get a hint that some quantum phenomenon is likely involved.) Then, we discuss two even more amazing experiments than EPR and simple proofs that no index-card theories can explain the results of those experiments. An index-card model for all of the quantum world is again shown to be *impossible*.

3.1 A Derivation of Bell's Central Inequality

We consider a set of n elements where each element is described in terms of *three* properties, a, b, and c. Each property comes in one of *two* types, denoted a^+, a^-; b^+, b^-; and c^+, c^-, respectively. To make this very concrete, we consider a fictional town and three *real* properties of its inhabitants.

3.1.1 A Fictional Town and Story[1]

Imagine a town of n people, where each person is described by the following three *real, genetic* properties:

[1] The idea of using a fictional town to explain Bell's inequality comes from [112], although a different town and story was used there.

Table 3.1 *The eight types of people, based on three binary properties. Made-up numbers of people of each type are shown on the right.*

a^+	b^+	c^+	7
a^+	b^+	c^-	4
a^+	b^-	c^+	9
a^+	b^-	c^-	2
a^-	b^+	c^+	1
a^-	b^+	c^-	5
a^-	b^-	c^+	8
a^-	b^-	c^-	6

a: The **a***bility* to *roll their tongue*, that is, to stick out their tongue in a tunnel shape: A person is type a^+ if they can roll their tongue and type a^- if they cannot.

b: **b***right sun sneezing*. A person is type b^+ if they sneeze when entering Bright sunlight from the dark and b^- if they do not.

c: **c***olor blindness*. A person is type c^+ if they *confuse* the colors red and green and c^- if they do not.

Each of these properties is called *binary* because each has exactly *two* mutually exclusive types.[2] Also, I have chosen actual *genetic* properties that are stable throughout one's life in order to emphasize a key point that will be made later.

With these three binary properties, there are *eight* possible *types* of people in the town. We show the eight types in Table 3.1. To complete the example, I *arbitrarily* (I swear) made up the number of people of each type. So, for example, in this fictional town of 42 people, there are 9 people who have the Ability to tongue roll (a^+), and don't sneeze in Bright sunlight (b^-), and are Color blind (c^+).

Now we make a numerical observation about certain types of people in the town:

Numerical Observation The number of people (11) who *can* tongue roll (a^+) and *do not* sneeze in bright sunlight (b^-), plus the number of people (9) who *do* sneeze in bright sunlight (b^+) and are *not* color blind (c^-), is greater than or equal to the number (6) of people who *can* tongue roll (a^+) and are *not* color blind (c^-).

[2] The phrase *mutually exclusive* means that if a person has one type of the property they cannot have the other type.

Check it out for yourself. Note the pattern of the property types mentioned: $a^+, b^-; b^+, c^-; a^+, c^-$. That is a hint of what is to come.

More generally The above observation is just a particular instance of a *general claim*, which we state formally in terms of abstract binary properties a, b, and c. We explain the notation we will use with two examples (you should be able to generalize from those): The number of people who have *both* property types a^+ and b^- is denoted by $\#(a^+b^-)$, and the number of people who have all three property types a^-, b^+, and c^- is denoted by $\#(a^-b^+c^-)$, and so forth.

3.2 A Bell Inequality

Bell's Central Inequality: In any *set* of people, and with any assignment of three binary properties a, b, and c to the people:

$$\#(a^+b^-) + \#(b^+c^-) \geq \#(a^+c^-).$$

Do try this at home. Make up your own imaginary town, with your own made-up numbers, and check it out.[3]

Now we prove that Bell's Inequality is true.[4]

3.2.1 Now We Prove It

Proof Certainly,

$$\#(a^+b^-c^+) + \#(a^-b^+c^-) \geq 0, \tag{3.1}$$

since both terms on the left side must be nonnegative.
Now add

$$\#(a^+b^-c^-) + \#(a^+b^+c^-) \tag{3.2}$$

to both sides. This yields

$$\begin{aligned} &\#(a^+b^-c^+) + \#(a^-b^+c^-) \\ &+\#(a^+b^-c^-) + \#(a^+b^+c^-) \\ &\geq \#(a^+b^-c^-) + \#(a^+b^+c^-). \end{aligned} \tag{3.3}$$

[3] Or even better, choose three binary properties you like, query the students in the next class you attend, and check whether the inequality holds (your instructor will appreciate the interruption).

[4] The proof given here is based on a proof developed by D. Harrison [103].

Notice that the two terms on the right side of inequality (3.3), (that is, after the "$\geq$" sign), both contain a^+ and c^-, but the first of those terms *also* contains b^-, while the second of those terms *also* contains b^+. So, the first term of (3.3) on its right side counts the number of people of type a^+c^- that also have property b^-, while the second term on the right counts the number of people of type a^+c^- that also have property b^+. But, property B is *binary* and its types, $b+, b^-$, are mutually exclusive, so those two terms must count *every* person of type a^+c^- exactly once. That is,

$$\#(a^+b^-c^-) + \#(a^+b^+c^-) = \#(a^+c^-).$$

Now apply the same reasoning to the first and third terms on the left side of inequality (3.3) and then again to the second and fourth terms on the left side of (3.3). The left side of (3.3) then cleans up to be

$$\#(a^+b^-) + \#(b^+c^-).$$

So,

$$\#(a^+b^-) + \#(b^+c^-) \geq \#(a^+c^-),$$

as claimed. Hence, Bell's Inequality is proved. ■

3.2.2 Nothing Tricky in This Proof

Notice that in this proof, we started with the assumption that the assignment of binary properties to the people was *arbitrary*, and never claimed any special properties of the numbers of people of any type. This means that in the town example, we could have picked any nonnegative integers to put in Table 3.1, and the inequality would have been correct for that assignment of property types.

Further, the only fact we used about the properties was that they were binary. Similarly, we only used simple arithmetic and simple intuitive logic in the proof.

Let's make that super explicit Let's make that point super, super (maybe *tediously*) explicit. In the first line of the proof, we used the property of arithmetic that when you add two numbers that are both greater or equal to zero, then the sum is also greater or equal to zero. In the second line, we used the property that when you add the same number to both sides of a correct inequality, you again get a correct inequality. Simple logic, along with the assumption that the properties are binary, was used in the next-to-last step of the proof, where we argued that

$$\#(a^+b^-c^-) + \#(a^+b^+c^-) = \#(a^+c^-).$$

The same logic was used in cleaning up the left side of inequality (3.3).

So, only simple arithmetic and only simple logical reasoning were used, along with the assumption that the three properties are binary. Hence, Bell's Inequality *must* be true for *any* set and *any* assignment of three binary properties a, b, and c to the elements of the set. And, this has nothing to do with quantum mechanics – it's a fact about sets in our everyday world.

Arithmetic is arithmetic no matter what set it is applied to. And the logic we used is so basic and intuitive that it should be unassailable. Again, we state for emphasis:

> Bell's Inequality is about arbitrary *sets* of elements where each element is described in terms of three binary properties. It has *nothing* to do with quantum mechanics or Einstein, or EPR experiments etc., or even weird towns. It is a universally true mathematical fact about sets and binary properties.

As pointed out in [112], Bell's Inequality could have been observed and proved by Aristotle (or any of the great ancient thinkers), but they would have had no motivation to do it – it would have seemed like a self-evident truth of no importance.

3.2.3 Connecting Bell's Inequality to EPR

Bell's Inequality is stated in terms of a set of elements where each element is described by three binary properties. So the inequality holds perfectly in the town story. But, our interest in Bell's Inequality is for use in recognizing when classical physics is not adequate to explain quantum phenomena, most clearly in the context of an EPR experiment. To address that question, we modeled classical physics (and hidden variables) using the index-card model. So, *under the index-card model* what are the binary properties and types in the context of an EPR experiment?

When we talked about the index-card model for an EPR experiment, we said that the particle "causes" a detector to flash Red or Green, depending on the index-card instructions the particle was given, and how the detector is oriented. So, in classical physics, under the index-card model of an EPR experiment, the three properties that a particle has are as follows:

a: What color light (Red or Green) a particle will cause the detector to flash when it reaches a detector with orientation A.

b: What color light (Red or Green) it will cause the detector to flash when it reaches a detector with orientation B.

c: What color light (Red or Green) it will cause the detector to flash when it reaches a detector with orientation C.

Then, under the index-card model, we can say that the particle has property type a^+, if when it reaches a detector oriented at angle A, the detector's light flashes Red; and that the particle has property type a^-, if when it reaches a detector at orientation A, the light flashes Green. Similar correspondences hold for property types b^+, b^-, c^+ and c^-, when a detector is oriented at angles B and C, respectively.

Now in each EPR trial, two *identical* particles are generated and given the same index-card instructions. So, under the index-card model, when two detectors have *different* orientations, they together should reveal two property types of the particle(s). For example, the joint property type a^+b^- means that one detector (it doesn't matter which) was oriented at angle A and flashed Red, and the other detector was oriented at angle B and flashed Green. Hence, under the index-card model, both particles should have properties a^+ and b^-. In particular, the particle going left out of the source should have those property types. In this view, each generated particle that goes left out of the source in an EPR experiment is analogous to a single person in the weird town story, and when the detectors have different orientations, the types of two different properties of the particle are revealed.[5]

Notice that each of the three terms in Bell's Inequality involve differing properties; a and b in the first term, b and c in the second term, and a and c in the third term. So, those EPR trials where the two detectors have different orientations provide the counts used in Bell's inequality. It follows that if, as is generally accepted, local realism is captured by the index-card model (via the concept of hidden variables), local realism leads to Bell's Inequality. That is, *if* the EPR experiment obeys local realism, the statistics obtained from a long EPR experiment *should* agree with Bell's Inequality.

3.2.4 And Yet!

Quantum mechanics predicts that Bell's Inequality does *not* always hold in EPR experiments. In one particular type of EPR experiment, the angle θ between orientations A and B is set equal to the angle between orientations B and C, so that the angle between A and C is twice that of the other two

[5] It may seem tiring and strange that phrases such as "under the index-card model" are used so often. This is done in order to be completely clear that we have been describing what the index-card model implies about the EPR experiment and the particles. We are *not* describing what quantum mechanics predicts, or asserting that the index-card model is a correct reflection of quantum reality.

angles.[6] In that case, for *any* θ strictly larger than zero and strictly less than 90 degrees, quantum mechanical predictions violate Bell's Inequality.

This gives another proof that no theory that obeys the index-card model can make all the same predictions that are made by quantum mechanics. And now, many experiments of this type confirm the correctness of quantum mechanical predictions that violate Bell's Inequality.

3.3 Why Is Bell's Inequality So Important?

Bell's Inequality applies to *any* set of elements with three binary properties. This makes it very easy to apply it to a wide range of situations or experiments. In particular, we can apply the inequality in experiments where a quantum phenomenon (such as entanglement) is suspected. And, reiterating, the importance of the inequality is that it is sometimes *violated.* That is a fact!

Really? But, how can that happen? The derivation of Bell's Inequality just relies on simple arithmetic and basic logic, so how could it *ever* be violated?

In truth, a large range of people (physicists, philosophers, physicians, (meta)physicians, philanthropists, philologists, philatelists, phrenologists, psychiatrists[7], and (amateur) phylogeneticists – like myself) continue to debate this, with no clear agreement.

Somehow, some of the assumptions that lead to Bell's Inequality must *not* hold in experiments where the inequality is violated. So, let's look at what assumptions lead to Bell's Inequality.

3.3.1 Local Realism Again

In the town story that illustrates Bell's Inequality, I chose three real properties (tongue rolling, sunlight sneezing, and color blindness) that are *genetic* and are *immutable* throughout one's life. The reason for choosing genetic properties is to *emphasize* that these properties *simultaneously* exist and persist whether anyone observes them or not, and no matter what other properties the inhabitants of the town may have, or what other properties are observed.

In fact, some people don't know that tongue rolling or sunlight sneezing is a thing, and many people who are color blind don't know it. But, these properties are real (Google them). So, they obey Einstein's concept of *realism.*

[6] A Bell inequality for this situation is derived in Exercises 15 and 16 at the end of the chapter. There, $\theta = 60$ degrees.

[7] I mean "phsychiatrists".

The genetic properties also obey Einstein's concept of *locality*. The property of being color blind, for example, can be determined by examining a specific person, and that property is not affected by, nor does it affect, any property of any other person, when *they* are examined.

The information that a particular person is color blind could be broadcast to everyone in the town, but it would not ever affect who else is color blind. So, using these genetic properties for illustration, there should be no doubt or confusion that Bell's inequality is correct and appropriate for the town.

So? In situations that obey local realism, Bell's Inequality *should* hold. This leads again to a conclusion that we saw in Chapter 2, that in the specific experiments where Bell's inequality is sometimes violated, the assumption of *realism* or *locality*, or both, do *not* hold. And, this again leads us to an *impossibility* theorem:

> No local realistic theory can fully explain the observations of experiments where Bell's Inequality is violated. And since quantum theory predicts such violations, no local realistic theory can agree with *all* predictions of quantum mechanics. Such a theory is *impossible*.

3.4 Is There a More Insightful Explanation?

When we see a violation of Bell's Inequality, it strongly suggests that quantum phenomena are present and that they are nonlocal, or that they do not obey the realism assumption.[8] But reality and locality are very high-level, subjective, somewhat philosophical concepts. Are there more concrete, less philosophical explanations or experimental results that more directly suggest why and when Bell's inequality will be violated?

3.4.1 Try This One

The Bell premise The underlying *premise* of Bell's inequality is that there is a set of elements characterized by three properties, a, b, and c, where each property has two possible types. And it is certain that each element does simultaneously have one type of each of the three properties. In fact, the purpose of

[8] But, there is even disagreement about the disagreement: "Most physicists take the point of view that it is the assumption of *realism* which needs to be dropped from our worldview in quantum mechanics ..." [79], and "The viewpoint of most physicists is that the violation of Bell inequalities shows us that quantum mechanics is *nonlocal*" [112], and "It seems to be a matter of changing fashion whether one blames locality or realism" [43].

telling the story of the weird town and the three genetic properties is to explicitly describe a situation where this premise is obvious and unquestioned.

From the premise, Bell's inequality is a purely mathematical statement about the number of elements with certain combinations of the three properties. Bell's inequality is correct for any set of elements that satisfies the underlying premise – it is as mathematically secure as any known math. But, in all of mathematics, any particular mathematical statement (e.g., a theorem) has premises (not always made explicitly) that must hold before we can apply that particular statement.[9] In order for Bell's Inequality to apply, the premise stated above must hold because Bell's Inequality refers to *pairs* of properties, selected from three properties, that elements simultaneously have. Consistent with this point, note that the proof of the inequality refers to combinations of all three properties that elements in the set have.

What does a "violation" imply? A situation where the observed numbers violate Bell's inequality, does not give evidence that Bell's mathematics is (sometimes) wrong – instead it just indicates that the premise of Bell's inequality does not hold in that situation – even though it may have looked at first like the premise does hold. And that is what is happening in the EPR experiment. From a mathematical perspective, explaining the "violations" is that simple and nonmysterious.[10] But, from the perspective of physics, we still want to understand why the premise does not hold in an EPR experiment.

The premise and EPR As detailed in Section 3.2.3, in the EPR experiment the three binary properties, a, b, c, are what color light (Red or Green) a particle causes a detector to flash when the particle reaches a detector in one of three orientations. Now, for each particle and each of the three orientations, one of the two colors *will* flash when the particle reaches the detector. Moreover, in a long-enough experiment, each color will flash in some trial(s), for each detector orientation; and the orientation of the detector is only chosen shortly before the particle reaches the detector.

So even though in a single trial, a detector only reveals the type of one property of the particle that reaches it, in a long-enough experiment, the collective results suggest that the particles possess one type of *each* of the three binary

[9] The conclusion of a particular theorem may be true even when the premises of that theorem do not hold, but we can only appeal to that specific theorem to explain the conclusions, when *its* premises hold.

[10] Also: "To avoid a common misstatement, we emphasize that Bell's *inequality* was violated [in EPR experiments]. Bell's ... derivation of the inequality from the assumptions of reality and separability [locality], is a mathematical proof not subject to experimental test." [91]

properties. This is what leads us to think that the premise of Bell's Inequality should hold in an EPR experiment. We are lulled into thinking that the particles form a set of elements which obey the underlying premise of Bell's inequality. And, if they do obey the premise, the observed numbers would absolutely agree with Bell's inequality.

But, the (three) properties that are measured in EPR experiments where Bell's inequality is violated are properties that, according to quantum theory, cannot be observed together in a single particle. Such properties are called *complementary*. In particular, quantum theory implies that we cannot observe what color light a particle will cause to flash if the detector is oriented vertically, and also observe what color light it will cause if a detector is oriented horizontally. For a single particle, we can learn either one of these things, but only one.[11]

Experiments And this is not just a prediction of quantum mechanics – it is something validated experimentally, as follows. The detectors used in an EPR experiment actually allow a particle to enter the detector, cause a color light to flash, and then exit the detector. In fact, the particle can go through several detectors arranged in series. So suppose we first arrange two detectors in series, both oriented the same way, say *vertically*. It turns out that the same color light will flash in both detectors, meaning that the property of the particle measured by the first detector will be unchanged by that first measurement, and will again be seen at the second detector. This makes it seem that the particle does inherently "have and hold" that observed property. Let's say that first observed property is a^+.

But, now we add a third detector, after the first two, and orient that detector *horizontally*. The color light that the third detector flashes will be *random*, unrelated to the color light flashed by the first two detectors. Let's say that the observed property is b^-. At that point, we are tempted to say that the particle has properties a^+ and b^-, that is, the particle is of type a^+b^-. We can test this expectation by adding a fourth detector, again oriented *vertically* (see Figure 3.1). If the particle *is* of type a^+b^-, the fourth detector should always

[11] You probably have heard of *Heisenberg's Uncertainty Principle*, which is often stated as something like:

> It is not possible to simultaneously determine, to arbitrary precision, both the position and the momentum (or velocity) of a particle.

Now, the properties of position and momentum are measured by numbers that are not always (or often) *discrete* (i.e., integers), while Bell's EPR experiments measure discrete events (e.g., which color light flashes), but the principle is the same – we can't determine the color light that a single particle will cause a detector to flash in more than one detector orientation.

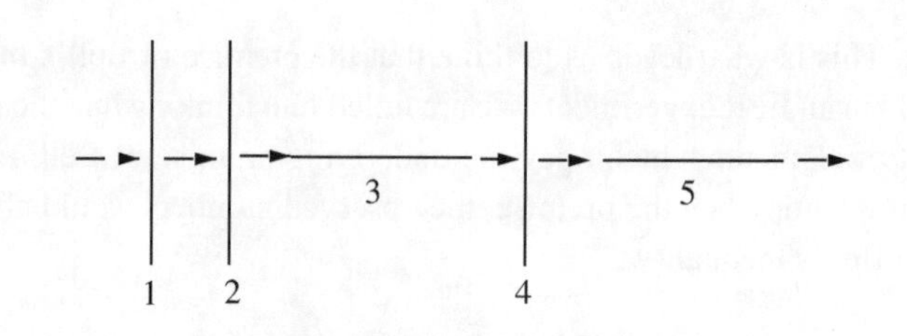

Figure 3.1 Detectors 1, 2, and 4 are oriented vertically; and detectors 3 and 5 are oriented horizontally. Detector 1 will randomly flash Red (or Green) with probability one-half. Detector 2 will flash the same color as Detector 1. Detector 3 will randomly flash Red (or Green) with probability one-half. The surprise is that Detector 4 will also randomly flash Red (or Green) with probability one-half, unrelated to what detectors 1 and 2 flashed. Similarly, Detector 5 will randomly flash Red (or Green) with probability one-half.

flash the same color light, a^+, that the first two (vertically oriented) detectors flashed. But it doesn't always flash the same color!

Now for the kicker: The color light flashed at the fourth detector will again be *random*, unrelated to the colors seen at the first three detectors. And, if we add a fifth detector oriented horizontally, the color flashed at that detector will also be random, unrelated to the previous colors – in particular, unrelated to the color flashed at the third (horizontally oriented) detector. So, we can't say that the particle *is* of type a^+b^-. We can only report what was observed at those five detectors.

What this means In the above experiment, we can't determine together the two properties of a particle. And, quantum mechanics says that properties of objects *do not exist* unless and until they *are* observed; it follows that quantum mechanics says that there *are no* particles that simultaneously have two, or all three, of the properties observable by EPR-type detectors. It isn't just that we cannot simultaneously *know* all the property types of a particle. It is that no particle can simultaneously *have* three, or even two, of the properties. According to quantum mechanics, such an object does not, and cannot, exist! In contrast, there can be a town citizen who simultaneously rolls their tongue, sneezes in strong sunlight, and is not color blind – and we can observe those properties all together. Quoting from [9]:

> The point is this. Quantum objects may in principle have a number of observable properties, but we can't gather them all ... in a single go, *because they can't all exist at once.*

So? Bell's inequality refers to properties of objects that *do* exist together, and can be measured together, as in our normal world and the weird town. But, when we try to reason about collections of (quantum) properties that *cannot* exist together in the quantum world, we get into trouble. Again, from [9], writing about Bell's Inequality:

> ... the problem is that we're assuming we can say something meaningful about a quantity that we don't measure [observe]. But ... we can only make meaningful statements about things that we *do* measure [observe] ... It is the inability to speak meaningfully about a quantity we don't measure [observe] that allows quantum mechanics to violate Bell bounds.

Well, that is the most meaningful explanation I have heard for *why and when* Bell's Inequality is violated. But, it is not completely satisfactory (to me), and so the violation of Bell's inequality in the quantum world remains somewhat magical and mysterious.

Mermin's take:

> The best explanation that anybody has come up with to this day is to insist that no explanation is needed beyond what one can infer from the laws of quantum mechanics. Those laws are correct. Quantum mechanics works. ... One gets puzzled only if one tries to understand how the rules [laws] work not only for the actual situations in which they are applied, but also in alternative situations that might have been chosen but were not. [74]

3.5 Using Impossibility to Create Possibility

There may not be a satisfying explanation for counterintuitive quantum phenomena, but with or without a good explanation, we can *use* Bell's Inequality to help recognize when quantum phenomena (particularly entanglement) might be present. And this has opened up many practical possibilities in quantum technology, as reflected in the earlier quote in Chapter 2:

> The quantum revolution that's happening now, and all these quantum technologies – that's 100% thanks to Bell's theorem.

One of the first people to see how to use Bell's inequality to create something that was not conceivable with classical physics was Artur Ekert [35]. As a young graduate student, he saw how violations of Bell's Inequality could be used to protect against eavesdropping on conversations between two parties. After a public lecture that Bell gave, Ekert had a very brief opportunity to explain the idea to Bell. As reported in [42], Ekert recalled that after enthusiastically telling Bell of the method, Bell stared at him and said:

> Are you telling me that this could be of practical use?

Ekert replied:

> Yes, I think it can.

And Bell said:

> Well, it's *unbelievable.*

Bell was then whisked away by other attendees. Bell died shortly after that and never got to see the blossoming of the possibility revolution that his impossibility revolution led to – but at least he did see its first germination.

3.5.1 But How?

There are several ways that Bell's inequality is used, and many of them follow the same general idea. In many applications, one needs to produce a large number of pairs of *entangled* particles whose "quantum values" are unknown and look randomly assigned (when examined). But, entanglement is sometimes hard to produce and maintain, so we also a need a way to be confident that the produced pairs are actually entangled.

To make this concrete, think of the (hopefully entangled) pairs of nickels coming out of the Philadelphia mint. The quantum value of a coin is whether it faces heads up or heads down when the coin is examined. If there are many pairs, and the statistics of the examined pairs violate Bell's Inequality, we can be very sure that the coins were entangled.

But, once a coin is examined, it can no longer be used in any application that requires *unknown, random* values – since a coin's value is known and fixed once it is examined. So how can we generate pairs of entangled particles and also test to be fairly sure that the particles are entangled?

The idea is simple: Generate many more pairs of (hopefully entangled) particles than are needed in the application. Then, *randomly* select many pairs to examine, leaving the number of unexamined pairs large enough for the application. If the number of examined pairs is large enough, and the statistics coming from the examined pairs violate Bell's Inequality, then we can be confident that the unexamined pairs are entangled. Those unexamined pairs are then used in the application.

3.6 GHZ: An Even More Magical Experiment and Amazing Theorem (with an Easy, Elementary Proof)

Bell's EPR-based theorems and inequalities are the most famous of all the mathematical results that make explicit the weirdness of entanglement and the difference between the quantum and the classical worlds. They are also the basis for most of the experimental tests for the reality of entanglement.

But, there are several additional theoretical ideas that have extended Bell's original work, and some of these have also been used in experimental tests of entanglement. The most amazing and famous of these extensions is called the *GHZ* experiment, named after its three developers, D. Greenberger, M. Horne, and A. Zeilinger.[12] The GHZ experiment was an unexpected consequence of the work that Zeilinger and colleagues were doing to develop *quantum teleportation*, where the state of one particle can be transferred to a distant particle without any direct interaction of the two particles (see [111, 112] for general audience expositions).

In the (first thought, and then real) GHZ experiment, instead of generating two entangled particles (as in EPR), *three* entangled particles are generated and shot out to three equally separated detectors. Each detector independently chooses to orient *vertically* or *horizontally* (relative to the incoming particle) just before a particle arrives. Also (to aid in the analyses), when a particle reaches a detector, instead of turning on a light, the detector either outputs the number $+1$ or the number -1. As before, we will not, and need not, explain the inner workings of the detector.

Similar to the case of the EPR experiment, we will see that no index-card model can make the same predictions that quantum mechanics makes about the GHZ experiment. The more amazing thing is that with a plausible restriction of the index-card model, we can prove that there are certain circumstances (which we detail below) where quantum mechanics predicts that a specific result of a GHZ experiment will occur with *certainty*, while under the index-card model, that result can *never* occur. These are completely opposite predictions, and notable because (unlike the case of EPR experiments) the predictions do not involve statistics or probability. Also, only a small number of experimental trials will likely be needed before seeing that result. And now the experiment has actually been done, although it is apparently quite difficult to create three

[12] The paper describing the theoretical analysis and proposed experiment [46] has an additional author, A. Shimony, who (I think) contributed to the theoretical analysis. David Mermin also had a hand in improving the idea, reducing the need for four entangled particles to three. Zeilinger shared the 2022 Nobel prize in physics for this and other related experiments in detecting and exploiting entanglement.

entangled particles. As expected, the experiments support the predictions of quantum mechanics.

3.6.1 Why This Matters

Recall that in the EPR experiment, quantum mechanics predicts that a specific outcome (agreement of the color of the lights) happens in *one-half* of the trials, while under the index-card model it would occur in at least *five-ninths* of the trials. The gap between one-half and five-ninths is large enough that when a large number of trials are run (say a million times, or even a few thousands), there is very little doubt about which prediction has been confirmed.

But, despite the experimental (statistical) validations of quantum mechanics based on Bell's Inequality, to further reduce any remaining doubt, we want experiments where the *gap* between the predictions (e.g., one-half vs. five-ninths) is as large as possible.

Or, even better, where quantum mechanics predicts that a specific result *must* happen, but any index-card theory predicts that it can *never* happen. Then, we don't have to rely on any statistical observations.

3.6.2 Bell's Theorem via the GHZ Experiment

Similar to Observations Q1 and Q2 stated in the discussion of the EPR experiment, in the case of the GHZ experiment, quantum mechanics makes the following two predictions, which have now been supported in real experiments:[13]

> **Quantum prediction GHZ1:** In any trial, when *all three* of the detectors are oriented *vertically*, an *even* number (0 or 2) of the detectors will output $+1$, and the other(s) will output -1. Hence the *product* of the numbers output in the trial will be -1.

> **Quantum prediction GHZ2:** In any trial, when exactly *one* of the three detectors is oriented *vertically* (and so the other two are oriented *horizontally*), an *odd* number (1 or 3) of the detectors will output $+1$, and any other detectors (two or zero of them) will output -1. Hence, the *product* of the numbers output will be $+1$.

These two predictions (now observations) are the only things we need to know about the outcomes of trials in a GHZ experiment. Note that these two predictions do not cover the cases that exactly *zero* or *two* of the detectors are oriented vertically, but those cases are not needed in the analysis. Note

[13] As before, we will not try to justify these predictions, as that requires hard-core quantum mechanics.

also that, unlike statement Q2 characterizing some of the behavior of the EPR experiment, neither prediction GHZ1 nor GHZ2 involves statistics or probabilities.

3.6.2.1 Can an Index-card Model Agree with GHZ1 and GHZ2?

As in our discussion of the EPR experiment, we ask: Can there be an *index-card* theory that predicts GHZ1 and GHZ2? More generally, can there be a local realistic theory that agrees with those quantum predictions?

Again, it is easy to create an index-card theory for each of the two predictions *separately*. In fact, it is trivial to do so. To satisfy Prediction GHZ1 by itself, just give all three particles the rule: "Always cause the detector to output -1". Then, none (0) of the detectors will ever output $+1$, which satisfies the requirements of GHZ1 but violates GHZ2. To satisfy Prediction GHZ2 by itself, just give all three particles the rule: "Always cause the detector to output $+1$." Then, all (3) of the detectors will output $+1$, which satisfies the requirements of GHZ2 but violates GHZ1.

It is possible to create less trivial rules that separately satisfy the two predictions, and look (for a short while) like they might satisfy both rules together. But, as we will next show, that is *impossible*.

3.6.2.2 The Answer Is No!

In the EPR experiment, we deduced that in any trial of the experiment, the two detectors must be given the same rules. But in the GHZ experiment, we cannot make that deduction, and different detectors can be given different rules. So, in any trial of the GHZ experiment, the particle going to detector 1 is given a rule that tells it which number ($+1$ or -1) to cause detector 1 to output if it finds the detector in the vertical orientation; and which number ($+1$ or -1) to cause detector 1 to output if it finds the detector in the horizontal orientation. In the same way, rules (which need not be the same as for particle 1) must be given to particle 2 going to detector 2, and rules must be given to particle 3 going to detector 3.

To specify the rules that particle 1 receives in a trial, we use *variables* $N(1, v)$ and $N(1, h)$, which take on values of $+1$ or -1. The values of these two variables specify the numbers that particle 1 will cause detector 1 to output, depending on the orientation of detector 1. The variable $N(1, v)$ specifies the number in the case that particle 1 finds detector 1 oriented vertically; and the variable $N(1, h)$ specifies the number in the case that it finds detector 1 oriented horizontally. Since there is no requirement that the particles receive the same rules, we will also have variables $N(2, v)$, $N(2, h)$, $N(3, v)$, and $N(3, h)$ to specify the rules given to particles 2 and 3. Note that the settings of these variables can be different in different trials.

When the particles leave the source, the orientations of the three detectors have not yet been chosen, so in an index-card theory, the rules given to the particles must cover *all* possible ways that the detectors could be oriented. So for any index-card theory, in each trial there are a total of six variables that will be given values (+1 or −1), and each particle will be told the values of two of those variables.

Can't get no satisfaction In order for an index-card theory to agree with Prediction GHZ1, in every trial the values given to the variables must satisfy the following equation:

$$N(1, v) \times N(2, v) \times N(3, v) = -1. \tag{3.4}$$

And in order to agree with Prediction GHZ2, in every trial the values given to the variables must satisfy the following three equations:

$$\begin{aligned} N(1, v) \times N(2, h) \times N(3, h) &= +1 \\ N(1, h) \times N(2, v) \times N(3, h) &= +1 \\ N(1, h) \times N(2, h) \times N(3, v) &= +1. \end{aligned} \tag{3.5}$$

Since each product in (3.5) must equal +1, their product must also equal +1. That is, in every trial, the variables must satisfy the equation (omitting the times symbol "×"):

$$N(1, v)N(2, h)N(3, h)N(1, h)N(2, v)N(3, h)N(1, h)N(2, h)N(3, v) = +1. \tag{3.6}$$

Collecting identical terms, this is equivalent to saying that in every trial:

$$N(1, v) \times N(2, v) \times N(3, v) \times N(1, h)^2 \times N(2, h)^2 \times N(3, h)^2 = +1. \tag{3.7}$$

Each of the variables $N(1, h)$, $N(2, h)$, and $N(3, h)$ must have value either +1 or −1, so each of the terms $N(1, h)^2$, $N(2, h)^2$, and $N(3, h)^2$ must equal +1. The product in (3.7) then simplifies to:

$$N(1, v) \times N(2, v) \times N(3, v) = +1. \tag{3.8}$$

But that contradicts (3.4), and hence is a *violation* of Prediction GHZ1.

Recapping: In order for an index-card model to satisfy Prediction GHZ2, the values assigned to the six variables must satisfy the three equations of (3.5) in every trial. But then, a little (plain-old, ignore-quantum-mechanics) arithmetic leads to equation (3.8) which violates Prediction GHZ1. Hence, we have the following version of Bell's theorem:

Theorem 3.6.1 *No theory that obeys the index-card model can agree with both quantum-mechanical Predictions GHZ1 and GHZ2. So, if every local-realistic theory can be modeled by an index-card theory, no local realistic theory can agree with all the predictions of quantum mechanics. Such a theory is impossible.*[14]

No inequalities, no statistics We pause here to state, again, a key difference between the Predictions GHZ1 and GHZ2, and Prediction Q2: Predictions GHZ1 and GHZ2 do not involve any statistics, and the proof of Theorem 3.6.1 also does not involve statistics or probabilities.[15] Hence, unlike the case of Bell's theorem for EPR experiments, where we need a (relatively) long experiment to distinguish the frequencies of one-half from five-ninths, a short GHZ experiment might suffice. That is, it is *plausible* that a GHZ experiment trying to verify that nature does not obey the index-card model (and hence cannot be fully explained by any local-realistic theory) might only require a few trials. We will shortly examine that issue and the related issue of *certainty*.

3.6.3 From First Principles

Before we can discuss real (not just thought) experiments, we need to re-prove Theorem 3.6.1 in somewhat different way.

Essentially, the proof we gave for Theorem 3.6.1 is a proof by *contradiction* that starts by *assuming* that there is an index-card theory that agrees with Prediction GHZ2, and then reaches a contradiction to Prediction GHZ1. But, now we want a more fundamental proof of Theorem 3.6.1, by first establishing inherent properties of any index-card theory for a GHZ experiment, without referring to any quantum mechanical predictions. This will be more than an academic exercise, as we will see when we discuss questions of *certainty* and *empirical observations* made in actual GHZ experiments.

Inherent properties of the index-card model for GHZ experiments The index-card model says that when the three particles leave the source, each

[14] A slightly more streamlined proof appears in [106] but originally comes from [70]. That proof is shorter, but I think the expanded proof given here is easier to follow.

[15] Mermin [70] notes that the lack of statistics or probabilities in the GHZ-based Bell theorem came as a surprise. Quantum mechanics is an inherently probability-based theory, so although Bell's EPR-based theorem was itself surprising, it was not surprising that a theorem used to distinguish classical physics (which involves *laws* and *certainties*) from quantum physics would involve probabilities. But, the Bell theorem based on the GHZ experiment does not involve probabilities or statistics. That adds to the mystery of quantum mechanics.

of the variables $N(1, v), N(1, h), N(2, v), N(2, h), N(3, v)$, and $N(3, h)$ has a *fixed* value, either $+1$ or -1. So, when the particles leave the source, any *product* of those values is fixed. Let's look at a particular product, dividing the variables into two bracketed subproducts, as follows:

$$\begin{aligned} &\Big[N(1, v)N(2, v)N(3, v)\Big] \\ \times &\Big[N(1, h)N(2, h)N(3, v)N(1, v)N(2, h)N(3, h)N(1, h)N(2, v)N(3, h)\Big] \\ &= N(1, v)^2N(2, v)^2N(3, v)^2N(1, h)^2N(2, h)^2N(3, h)^2. \end{aligned} \tag{3.9}$$

We can now see that no matter how the variables are set, the product in (3.9) must be equal to $+1$, since each of the variables is squared (i.e., each variable occurs exactly twice in the product), and each variable must have value either $+1$ or -1. Note that this fact is deduced from the index-card model alone, and makes no reference to GHZ1 or GHZ2 or any other fact about quantum mechanics.

Further deductions We can make further deductions from the fact that the product in (3.9) is equal to $+1$.

If the variables $(N(1, v), N(2, v), N(3, v))$ in the first bracketed subproduct in (3.9) have values where their product is $+1$, then the variables in the second bracketed subproduct must have values where their product is also $+1$.

Conversely, if the variables in the first bracketed subproduct in (3.9) have values so that their product is -1, then the variables in the second bracketed subproduct must have values so that their product is also -1.

Summarizing, in any index-card theory, in every trial, the two subproducts in (3.9) are *always equal*, either both $+1$ or both -1. Again, this deduction is made without reference to quantum mechanics – it is a property inherent in the index-card model for the GHZ experiment.

Theorem 3.6.2 *In any index-card theory, the two bracketed subproducts in (3.9) must be* equal *in every trial of a GHZ experiment.*

Do the deductions agree with GHZ1 and GHZ2? Of course not! If an index-card theory agrees with GHZ1, then (as stated in (3.4)), the first bracketed subproduct in (3.9) must equal -1. Now, view the second bracketed subproduct in (3.9) as

$$\big[N(1, h)N(2, h)N(3, v)\big]\big[N(1, v)N(2, h)N(3, h)\big]\big[N(1, h)N(2, v)N(3, h)\big].$$

If an index-card theory agrees with GHZ2, then (as stated in (3.5)), each of these three subproducts must be equal to $+1$, so the second bracketed subproduct in (3.9) must be $+1$. These two facts contradict Theorem 3.6.2, and this gives us a more fundamental proof of a Bell theorem:

Theorem 3.6.3 *In the context of GHZ experiments, no index-card theory can agree with all predictions made by quantum mechanics.*

3.6.4 Certainty and Experiment

Recall the statement made on page 54 that we want an experiment where there are circumstances such that "quantum mechanics predicts that a specific result *must* happen, but any index-card theory predicts that it can *never* happen." The GHZ experiment can satisfy this desire. As explained in [102]:

> ... the [EPR-based] test of Bell's theorem [Theorem 2.2.1] gives a circumstance where the quantal [quantum-mechanical] probability of something [specific] happening is 50%, while the local deterministic [local realistic] probability is more than 55%, [but] the GHZ variation gives a circumstance in which the quantal probability [of a particular result] is 1 and the local deterministic probability [of that result] is 0. [102]

In this quote, the "specific happening" referred to is that the two lights agree in color. The "particular result" referred to is discussed next.

A confession: The only way I understand to validate the claim that "... the GHZ variation gives a circumstance in which the quantal probability is 1 and the local deterministic probability is 0" requires making an additional assumption in the index-card model. The additional assumption is also needed in order to argue that in a real GHZ experiment, only a small number of trials will likely be required.

> **The deterministic index-card assumption** When a GHZ experiment begins, the values of the six variables $N(1, v)...N(3, h)$ are set and fixed for the entire experiment (or at least for, say, twenty trials). That is, values of the variables must be the same in each of those trials. We call this the *Deterministic Index-Card Assumption.*

Recall that we did *not* need this assumption in order to prove Theorem 3.6.1. The Deterministic Index-Card Assumption is a limitation on our prior index-card model, but is not unreasonable in that the index-card model is intended to capture the behavior in our normal, classical-physics, *deterministic* world. The word "deterministic" means that when the conditions (e.g., the behavior of the source and the settings of the detectors) in an experiment are the same in two different trials, the outcomes of the trials should be the same. This is an assumption made in classical physics, and I believe that this assumption is made (implicitly) in the GHZ analyses in [70] and [46].

The Deterministic Index-Card Assumption may fail as the experiment proceeds (say because the equipment heats up, or wears out, or a stray cosmic ray passes by, etc.), but it is plausible that this failure will not happen quickly.

Consequences For a real GHZ experiment, we define $R(1, v)$ to be a variable whose value, $+1$ or -1, is the most recent output of detector 1 when it was oriented vertically.[16] We similarly define the variables $R(1, h)$, $R(2, v)$, $R(2, h)$, $R(3, v)$, and $R(3, h)$ to record those (most recent) outputs.

Now suppose that an index-card theory agrees with (real) experimental outcomes, so that $N(1, v)$ equals $R(1, v)$ every time detector 1 is set vertically. A consequence of the Deterministic Index-Card Assumption is that $R(1, v)$ *must* remain constant throughout the experiment since $N(1, v)$ is constant. Similarly, if an index-card theory agrees with experimental outcomes, then $N(1, h) = R(1, h)$; $N(2, v) = R(2, v)$; $N(3, v) = R(3, v)$; and $N(3, h) = R(3, h)$; and the values of those N and R variables remain unchanged after they are first set in the experiment.[17]

In an experiment After we observe the values of $R(1, v)$, ..., $R(3, h)$ in an experiment, we can check whether the two subproducts

$$\left[R(1, v)R(2, v)R(3, v)\right] \tag{3.10}$$

and

$$\left[R(1, h)R(2, h)R(3, v)R(1, v)R(2, h)R(3, h)R(1, h)R(2, v)R(3, h)\right] \tag{3.11}$$

are equal or not.

The assumption that each of the six R terms is equal to its corresponding N term, and Theorem 3.6.2, means that any index-card theory must predict that the observed subproducts in 3.10 and 3.11 will be *equal*. But, by GHZ1 and GHZ2, quantum mechanics predicts that the two subproducts will be *unequal*.

[16] For clarity, note that the *definition* of $R(1, v)$ allows its value to change over the life of an experiment.

[17] So, if one of these six R values ever does change during an experiment, then we have direct evidence that no index-card theory (obeying the Deterministic Index-Card Assumption) can agree with the experimental outcome.

Thus, the "particular result" referred to earlier is that "the observed subproducts in 3.10 and 3.11 are unequal." So, we have now validated the earlier quote that there are circumstances where the "quantal probability [of a particular result] is 1, while the local deterministic probability is 0", showing the surprising fact that differences between classical and quantum-mechanical physics do not always need to involve probabilities.

How long does it take? How many trials are expected before the values of all six R variables are known? On the first trial, three of the six R values are revealed; and in subsequent trials, the settings of the detectors are equally likely, and are independent of prior trials. So, it should be intuitive that the expected number of needed trials to reveal all six R values is fairly small. It is beyond the scope of this book to do the analysis from probability theory, but the average number of needed trials (if we do the experiment repeatedly to see how many trials it takes before all of the R values are set, and then take the average of those numbers) is just *four*. So, if we budget say 20 trials, we will almost certainly observe all the six R values within those trials, and can then compare the two products. Hence, assuming the Deterministic Index-Card Assumption holds until all six R values have been seen, we conclude:

Theorem 3.6.4 *After a small number of trials of the GHZ experiment, we will almost surely have a circumstance where quantum mechanics makes a prediction that is the* opposite *of the prediction made by any index-card theory.*

3.6.5 Another Confession

I have to admit that in the literature on the GHZ experiment, statements stronger than Theorem 3.6.4 have been made. For example (translating statements in [69] into the terminology used in this chapter):

> If index-card rules existed, then trials where all detectors are oriented vertically would *always* have to produce an odd number of +1 outputs (i.e., the opposite of Prediction GHZ1). But they never do ...
>
> Thus a *single* trial where all detectors are vertically oriented suffices by itself to give data inconsistent with the otherwise compelling inference of index-card rules. (italics are original)[18]

[18] The actual statements are: "If the instruction sets existed, then 111 runs would *always* have to produce an odd number of red flashes. But they never do ..." and "Thus a *single* 111 run suffices by itself to give data inconsistent with the otherwise compelling inference of instruction sets." (italics are original).

This claim is essentially the claim that any index-card theory will always fail to agree with GHZ1, rather than what we have proved – that any index-card theory will either fail to agree with GHZ1 or with GHZ2.

I do not understand the reasoning behind those quoted statements. Still, even without those statements, what we have established here (Theorem 3.6.4) is strong enough for the most important conclusions about the GHZ experiment: (1) No index-card theory can agree with all quantum mechanical predictions about GHZ; (2) there are circumstances where quantum mechanics predicts that a specific result *must* occur, while every index-card theory predicts that it can *never* occur; and (3) only a few real experimental trials are likely needed to reveal whether that result does or does not occur.

3.7 Surprise Bonus: The Hardy Experiment and an (Almost) Trivial Proof of Bell's Theorem

Lucien Hardy [49] devised an experiment in 1992 that leads to the simplest known proof of Bell's theorem, that is, that no theory that obeys the index-card model can make all the same predictions that quantum mechanics makes.[19] Mermin [72] states that Hardy's experiment and analysis

> ... constitutes the simplest version of Bell's Theorem that I can imagine.

This further illustrates the major premise of this book that simpler proofs of impossibility theorems have become known over time and can be exposed to a large lay audience.

3.7.1 The Experiment

Similar to the EPR experiment, in Hardy's experiment there is again a central source that, in each trial, shoots two particles in opposite directions toward two detectors. Also, as in EPR, in each trial, a detector either flashes a *Red* light or a *Green* light. But now, the detectors can only be set in *two* (rather than three) possible ways, which we will just denote as A and B. As before, we do not need any details of the innards of the source or the detectors, and explaining those

[19] I knew about the existence of the Hardy experiment before writing this book, but I hadn't looked into the details because it gets much less press than the EPR or GHZ experiments. But, during a final review of this chapter, I learned that David Mermin wrote an exposition of the experiment and proof [72] in his trademark way – getting to the heart of the analysis without first discussing the complex quantum-mechanical details. So, of course, I had to read what he wrote and write what I learned.

details would involve hard-core quantum mechanics, which is way outside the scope of this book. All we need to know is that in this experimental setup, quantum mechanics predicts the following three things:

(HQ1) When the detectors are set differently (i.e., one to setting A and the other to setting B), it is impossible for both detectors to flash *Green*.
(HQ2) When both detectors are set to B, sometimes both detectors flash *Green* (about 9% of the time when both detectors are set to B, but we won't need this number).
(HQ3) When both detectors are set to A, it is impossible for both to flash *Red*.

Remember that with the index-card model, rules must be specified in each trial that state exactly what light each detector will flash, depending on the detectors' settings. In the EPR experiment, we deduced that in any trial, the rules for the two detectors must be the same (in order to satisfy prediction $Q1$), but that constraint does not follow from the three predictions in the Hardy experiment. So, each rule needs to specify a detector, a setting, and a color. We write each rule by specifying those three components in that order. For example, rule 2 (B, G) means that if detector 2 is set to B, then it must flash Green. Hence, in each trial, four rules must be specified to cover the two ways that each of the two detectors could be set.

Theorem 3.7.1 *No theory that obeys the index-card model can satisfy predictions HQ*1*, HQ*2*, and HQ*3*.*

Proof Suppose that there is an index-card theory that satisfies prediction *HQ*2. Since both detectors sometimes flash Green when they are both set to B, there must be trials where rules 1 (B, G) and 2 (B, G) are included in the rules given to the two particles. We call those the *BG* trials. Note that this only specifies something about rules, not anything about the actual way that the detectors are set. So, there can be (and almost surely will be) *BG* trials where a detector is not set to B.

Say whaaat? What can we say about the other index-card rules in *BG* trials? The rule 1 (A, G) can't be assigned in a *BG* trial. If it were, then it could happen that detector 1 is set to A and hence flashes Green, while detector 2 is set to B and also flashes Green. But those settings and outcomes violate HQ1. By the same reasoning, rule 2 (A, G) can't be assigned in a *BG* trial. So, the rules 1 (A, R) and 2 (A, R) must be included in all *BG* trials. But then if both detectors are set to A in a *BG* trial, both will flash Red, violating HQ3. ■

> Thus the regularities described in [HQ1, HQ2, HQ3] are inconsistent with a very common-sense – indeed, an apparently unavoidable – explanation for them, leaving those correlations profoundly mysterious. [72]

3.8 What Impossibility Proves

John Bell is widely quoted as having written:

> What is proved by impossibility proofs is lack of imagination. John Bell (1982)

Ouch!! When I first saw this quote I almost had a panic attack. Here, I was writing a book on impossibility proofs, one of the main ones being by John Bell, and in this quote he seems to be telling us to be skeptical about impossibility. He is not telling us to disbelieve his *proofs*, but maybe to disbelieve their *importance*.

What to do? I could ignore it, but that seemed dishonest, and I also thought that I must be missing the point somehow. I mean, John Bell, the author of the two most important impossibility theorems in physics (only one of which we discuss in this book), is telling us that they just prove a "lack of imagination." I could not simply ignore that – I had to understand it.

At first, I thought that maybe this quote was a not-too-subtle swipe at John von Neumann, who had written, in 1932, a "proof" that hidden-variable theories of quantum mechanics are impossible. But several people, the mathematician Grete Hermann in 1935, and even Einstein (as reported in [42]), noted a mathematical *assumption* in von Neumann's proof that has been characterized as "absurd" [15] and "silly" [71] and certainly unjustified, in the context of quantum mechanics, severely limiting the scope and relevance of what von Neumann proved.

Independently, Bell saw this error, although many years later, because Hermann's observation was not widely known. It seems that

> ... nobody listened to her, partly because she was an outsider to the physics community – and partly because she was a woman. [11]

Also, until 1955, the von Neumann proof had only been available in German, which Bell could not read.[20] Moreover, David Bohm had developed a (nonlocal) hidden-variable theory of quantum mechanics that did what von

[20] Why didn't he just use Google translate? Probably because it doesn't handle mathematical symbols well.

Neumann thought he had proven was impossible. As Bell said about Bohm's theory in [15]: "In 1952, I saw the impossible done."[21]

So, I thought that maybe Bell's statement was directed at the von Neumann part of the story. Indeed, maybe that is part of the explanation, but it is not the whole story. Again, it was David Mermin who clarified things, this time in [71]:

> John Bell did not believe that either of his no-hidden-variables theorems excluded the possibility of a deeper level of description than quantum mechanics, any more than von Neumann's theorem does. He viewed them all as identifying conditions that such a description would have to meet.

So, it was John Bell the *physicist* telling John Bell the *mathematician* that there may be deeper physical theory that again violates the mathematical constraints identified by his theorems, and yet validates Einstein's (and Bell's) intuitive view of physical reality. John Bell was saying that John Bell's impossibility theorems must not discourage people from looking for that deeper *physical* theory. And, maybe Bell was just being a bit too much of a contrarian, even being contrary to himself. He was famous for his contrarian personality. Quoting again from [71], writing about Bell:

> ... I ... insist that he is unreasonably dismissive of the importance of his own impossibility proofs. One could make a complementary criticism of much of contemporary theoretical physics: What is proved by *possibility* proofs is an *excess* of imagination. Either criticism undervalues the importance of defining *limits* to what speculative theories can or cannot be expected to accomplish. (italics added)

3.8.1 Final Praise

The impossibility proofs in this and the previous chapter are so simple that one might lose sight of the stellar accomplishments of *devising* experiments (EPR, GHZ, Hardy) where the quantum-mechanical predictions lead to such simple proofs. And of course, we must always appreciate the genius of John Bell in imagining that such proofs were possible, and in finding the first ones.

[21] I was confused by this when I first learned about it. I thought that Bell's theorem implies that hidden-variable theories of quantum mechanics were impossible, but Bohm's theory is a hidden-variable theory. The resolution is that Bell's theorem implies that no *local realistic* hidden-variable theory of quantum mechanics is possible. Bohm's hidden-variable theory is *nonlocal*. Von Neumann's impossibility theorem implies that even nonlocal hidden-variable theories are impossible, but remember that von Neumann's proof is not applicable due to an inappropriate mathematical assumption he made.

3.9 Exercises

1. The town story (as given above) illustrates Bell's inequality, but the story does not fully align with how EPR experiments are actually done. We want to use the story of the town, where Bell's Inequality is clearly appropriate and meaningful, to explain what we should expect in an EPR experiment *if* there are no quantum phenomena. So, we have to modify the town story.

So far, our town could be any town (even Champaign, Illinois, or Davis, California). It was not special. Now we are going to consider a special kind of town. This town is made up of *identical twins*.[22] In this town, we could again compile a table like Table 3.1, showing the number of people of each type, but now, because everyone is an identical twin, every number in that table would have to be even.

The story has gotten more complicated, but there is one key point:

> The fact that the people in this town are made up of identical twins does *nothing* to change Bell's Inequality.

Explain this statement.

If we want to explicitly reflect the modified town story, we can change Bell's Inequality to:

> **Proportional Bell Inequality** The *proportion* of *twin pairs* of type a^+b^-, plus the proportion of twin pairs of type b^+c^-, is greater than or equal to the proportion of twin pairs of type a^+c^-.

Explain why this is correct.

2. **Another twist to the story** Up until now, we just said that each person in the town is described in terms of three binary properties, *a*, *b*, and *c*. But, we didn't discuss *how* these properties are determined. Now we add the following details:

> The town has two examiners who (for now) determine the types of *all three* properties of each person in the town. For example, an examiner may look at a person and determine that they can roll their tongue, that they sneeze in bright sunlight, and that they aren't color blind, that is, they are type $a^+b^+c^-$. Hey, that's me!
> Further, for every pair of twins, the individual twins are examined *separately* in distant locations, by different examiners, at roughly the same time. Also, the twins

[22] There is an ambiguity in English when discussing twins, so let me state that when I use the word "twin" I mean one person, and when I use the phrase "pair of twins" I mean two people who happen to be twins of each other. I will sometimes just say they form a "pair of twins", or are a "twin pair". The phrase "pair of twins" never refers to four people.

(and the two examiners) are not permitted to communicate with each other until the examinations of that pair of twins are finished.

Any change yet? Does this added detail about the town and its examiners change Bell's Inequality or the Proportional Bell Inequality? Explain.

3. **One final twist to the story** Now we make one final change to the town and the story, which *will* actually modify Bell's Inequality, but the essential meaning and impact of the modified inequalities will be the same as before.

In this modification, an examiner will only look at a *single* property of the twin they are examining. So now, the examiner does not determine the types of *all three* properties of a twin, as was the case before, but only determines the type of *one* property. Which property, however, is selected *randomly*, each time a twin is examined?

In particular, when an examiner examines a twin, they first roll a three-sided die (or flip a three-sided coin if they have one) to *randomly* decide which particular property, a, b, or c, to look at. *Independently*, the other examiner, who is examining the other twin in the twin pair, also rolls a three-sided die to randomly decide which property to look at.

Because of these random selections over a large number of individuals (say a million, for a large town), an examiner will look at property a about *one-third* of the time, they will look at property b about one-third of the time, and will look at property c about one-third of the time.[23]

Now we ask: How *often* will one examiner look at property a in one twin, and the other examiner look at property b in the other twin?

And, because the twins are identical, when one examiner looks at property a and the other looks at property b, the effect is that the a type and the b type of *both* twins will be determined.

Equivalently: How *often* will *both* properties a and b be examined for a twin pair?

4. **And answer:** The answer is about *two-ninths* of the time. That is, the two examiners will together look at both properties a and b in about two-ninths of the twin pairs.

Explain why this is correct. Hint: It may help to give the examiners distinct names, say "Alice" and "Bob". Then, there are two ways that one examiner looks at property a in one twin and the other examiner looks at property b in the other twin: either Alice looks at property a and Bob looks at property b, or Alice looks at property b and Bob looks at property a.

[23] The phrase "about one-third of the time" is vague, but in fact if a million people are examined, and the property looked at is randomly decided each time, the deviation from one-third will be *extremely* small.

5. **And:** How often will both properties b and c be examined for a twin pair? Same question for a and c.

6. **Next question** Is it true that each of the three *pairs* of properties included in the Bell's Inequality will be examined roughly the *same* number of times?

7. **What about the proportions now?** If we determine the ab property type of *every* twin pair in the town, then we will know exactly how many twin pairs are of type a^+b^- (and will know how many twin pairs there are in the town), so we can divide those two numbers to get the proportion of twin pairs who are type a^+b^-.

But, in the last twist of our town story, we don't determine the a and b types of every twin pair.

So, how can we be confident that we can correctly determine what (rough) proportion of twin pairs are of type a^+b^-, as is needed in the Proportional Bell Inequality? We can ask the same question about the other two property types included in the Proportional Bell Inequality.

8. **A bit about sampling and large numbers** Try the following experiment. Put 100 coins into a box and shake the box so that the sides of the coins showing heads or tails are well mixed. Then, randomly reach in and select *two-ninths* of the coins (about 22), without flipping over any selected coin. Record how many of the selected coins are heads up and how many heads down.

What do you conclude from this experiment?

If you don't want to collect 100 coins, just flip a coin 100 times, and write out the results in the order they occur. Then, randomly look at any 22 consecutive results and determine the proportion of times heads were up and the proportion of times heads were down in those 22 results.

Now imagine that instead of 100 coins (or flips) you have *one-million* coins, and that instead of randomly selecting 22 coins, you select about 220 *thousand* coins. And again, you see how many selected coins are heads up or heads down.

What do you think the results will be now, compared to the results with only 22 selected coins out of 100 coins?

9. **Two states good, four states better** The above experiment shows the power of *sampling* and of *large numbers*, to compute the proportion of objects (coins) of one type (heads up) versus another type (heads down). But in the town story, when one examiner looks at property a and other looks at property b, there are *four* possible results, not just *two*. Explain.

So, if we examine 222,000 twin pairs out of one million pairs, can we be confident that we can correctly determine what (rough) proportion of twin pairs are of type a^+b^-, as is needed in Bell Inequality 2? Hint: Yes, we can! Explain.

10. **Finally:** So, with the final twist to the town story, we now conclude:

> **Third Bell Inequality** In a large town (say one million pairs of twins), the proportion of twin pairs *determined* by the examiners to be type a^+b^-, plus the proportion of twin pairs *determined* to be type b^+c^-, is greater or equal to the proportion of twin pairs *determined* to be type a^+c^-.

Explain why this is correct.

11. **Back to EPR** Now that we have filled out the town story, fully explain how it relates to an EPR experiment, and why Bell's inequality should hold in an EPR experiment *if* no quantum behavior occurs in the experiment.

The answer to the first part of this question is essentially given in Table 3.2, but don't look at it until you answer the question yourself.

12. **Symmetric Bell Inequality**

Bell's Inequality is:

$$\#(a^+b^-) + \#(b^+c^-) \geq \#(a^+c^-).$$

We can similarly derive a Symmetric Bell inequality:

$$\#(a^-b^+) + \#(b^-c^+) \geq \#(a^-c^+).$$

The Symmetric Inequality comes from interchanging all the superscripts "−" and "+" in Bell's Inequality. You can argue that the Symmetric Inequality is correct by appeal to the concept of *symmetry*. If you are comfortable with that, make that argument explicitly. Otherwise, go through the proof of Bell's Inequality given in this chapter and make the needed changes, line by line, to prove the Symmetric Bell Inequality.

13. **Check it out** Use the numbers in Table 3.1 to verify that the Symmetric Bell Inequality holds for those numbers. I again swear that I have not rigged the numbers to make this work out. In fact, I am so confident of the outcome, I will not even check it before publication. (Yeah, right!)

14. **An added Bell Inequality** Now add Bell's Inequality to the Symmetric Bell Inequality. This yields the *Added Bell Inequality*:

$$\#(a^+b^-) + \#(a^-b^+) + \#(b^+c^-) + \#(b^-c^+) \geq \#(a^-c^+) + \#(a^+c^-). \quad (3.12)$$

Table 3.2 *Correspondence between a generic EPR experiment and the silly town story.*

EPR experiment	Town story
A *source* repeatedly generates two *identical* particles.	The town consists of *identical* twins.
When the source generates two identical particles, they are shot out in opposite directions each toward a *detector* on its side.	Two *examiners* are at distant locations, and each twin in a twin pair is examined by a different examiner.
Just before a particle reaches a detector, the detector will choose with equal probability to orient at one of three angles (A, B, or C).	When an examiner examinesa twin, they choose at random one of three binary properties (a, b, or c) to examine.
At roughly the same time, one particle in a pair reaches one detector and the other particle reaches the other detector.	At roughly the same time, one examiner looks at one twin in a twin pair, and the other examiner looks at the other twin.
When a particle reaches a detector, the detector either flashes a Red or a Green light, and reports its angle (A, B, or C) and the color (Red or Green) of its light.	When a twin is examined, the examiner reports one of the types: a^+ or a^-, b^+ or b^-, c^+ or c^-, depending on the property examined.
Compute three proportions from all the data *reported* by the detectors:	Compute three proportions from all the data *reported* by the examiners:
1) The proportion of particle pairs where one detector reports angle A and color Red, and the other detector reports angle B and color Green.	1) The proportion of twin pairs where one examiner reports a twin has property type a^+ and the other examiner reports that the other twin has property type b^-.
2) The proportion of particle pairs where one detector reports angle B and color Red, and the other detector reports angle C and color Green.	2) The proportion of twin pairs where one examiner reports property type b^+ and the other examiner reports c^-.
3) The proportion where one detector reports angle A and color Red, and the other reports angle C and color Green.	3) The proportion where one examiner reports a^+ and the other reports c^-.

I could ask you to use Table 3.1 to verify that the Added Bell Inequality holds for those numbers. But, given the material in this chapter, including Exercise 13, why is an explicit check of the table unnecessary? That is, how can I now really be sure it will all work out without actually checking?

15. **Constrained EPR** In the generic EPR experiment, either detector 1 or detector 2 is allowed to choose any of three prespecified orientations A, B, or

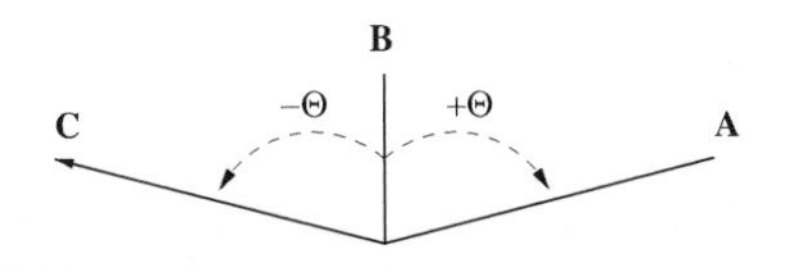

Figure 3.2 The three allowed orientations of a detector as seen from the side of the EPR experimental setup. See Figure 3.3 for a complete enumeration of how the orientations of the two detectors can differ.

C. But suppose, as was done in Section 3.2.4, that we constrain the orientations so that a detector can only be oriented *vertically* (orientation B), or at some prespecified angle $+\theta$ in the clockwise direction from vertical (orientation A), or at angle $-\theta$ in the counterclockwise direction from vertical (orientation C).

From the perspective of an observer off to one side of the experimental setup, we label these orientations as $A, B,$ and C going right to left (see Figure 3.2). So, although the two detectors are separated, the angle between their orientations is always *zero*, θ or $2 \times \theta$.

Look back at Figure 2.1. In that figure, the experimental setup is shown from the perspective of an observer off to one side of the experiment. From that perspective, there are *six* ways that the two detectors can be oriented differently (and of course, three ways that they can be oriented the same). See Figure 3.3.

But, does the universe know its left from its right? The universe has no up, down, right, left, in, out, north, south, vertical, horizontal, clockwise, counterclockwise, etc.[24]

Use this observation about the universe to argue that when the detector orientations are constrained as assumed, $\#(a^+b^-)$ should roughly equal $\#(b^+c^-)$, and that $\#(a^-b^+)$ should roughly equal $\#(b^-c^+)$ in a long EPR experiment. Explain fully that in a long EPR experiment of N trials, the following should hold:

$$2 \times \left[\frac{\#(a^+b^-)}{N} + \frac{\#(a^-b^+)}{N}\right] \geq \frac{\#(a^+c^-)}{N} + \frac{\#(a^-c^+)}{N}. \tag{3.13}$$

16. **Agree to disagree** Recall that when two detectors flash different colors, those detectors are said to *disagree*. The proportion of trials where detectors disagree is called the *disagreement rate*. Use the observation about the universe to conclude that the disagreement rate, in a long EPR trial, should just depend on the angle θ between the two detectors and not on their exact orientations.

[24] And by the way, when you are at the North pole, which way is North?

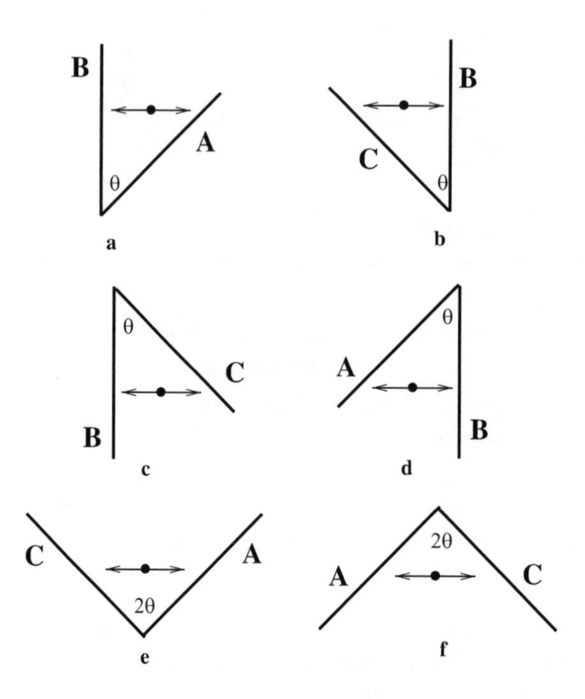

Figure 3.3 Looking from the side of the experimental setup, there are six ways that the two detectors can be oriented differently. The point and double arrow represents the source between the two detectors.

With that conclusion, argue that the right side of (3.13) is the disagreement rate of two detectors oriented with angle 2θ between them; and the left side of (3.13) is twice the disagreement rate of two detectors oriented with θ degrees between them.

Summarizing, we now have an

Angled Bell Inequality:

Twice the disagreement rate, when two detectors are separated by an angle θ, must be *greater than or equal to* the disagreement rate when two detectors are separated by angle 2θ.

The Angled Bell Inequality is another well-known and well-used version of Bell's Inequality. See [91] for a more physical, less mathematical explanation of this inequality.

Just for those who are curious, for the EPR experiment assumed in this chapter, quantum mechanics predicts that the disagreement rate is $\sin^2(\theta/2)$.

For example, when $\theta = 60$ degrees, quantum mechanics predicts that the disagreement rate will be $\frac{1}{4}$, so twice that is $\frac{1}{2}$. For $\theta = 2 \times 60 = 120$ degrees, quantum mechanics predicts that the disagreement rate will be $\frac{3}{4}$. So, when $\theta = 60$ degrees, the predictions of quantum mechanics violate the Angled Bell Inequality, since $\frac{1}{2} < \frac{3}{4}$.

17. **The GHZ theorem** In the development of Theorem 3.6.1, we started with the assertion that any index-card model must be consistent with Prediction GHZ2, and so must obey the three equations in (3.5). From that assertion, we deduced the equation in (3.8), which violated Prediction GHZ1.

An alternative approach is to start with the assertion that any index-card model must be consistent with Prediction GHZ1, and so the product $N(1, v) \times N(2, v) \times N(3, v)$ must equal -1. Finish that approach to proving that no index-card model can agree with all the Predictions of quantum mechanics.

18. In our discussion of the EPR experiment, we first deduced that any index-card model must assign, in each trial, the same rules to the two particles. In the GHZ experiment, we did not make that assumption nor make that deduction. For example, $N(1, v)$ need not be equal to $N(2, v)$ or $N(3, v)$ in any trial. But, suppose that we did add the restriction to the index-card model that the particles must be given the same index-card rules in any trial – the rules can change between trials, but in each trial, each particle is assigned the same rules.

First, explain why this restriction does not invalidate the conclusion that the index-card model is incompatible with quantum-mechanical predictions GHZ1 and GHZ2?

Second, does this restriction make it easier to deduce that such index-card rules are incompatible with GHZ1 and GHZ2?

19. Why do we need the deterministic index-card assumption in order to derive Theorem 3.6.4?

20. **The CHSH inequality** A variant of Bell's inequality was derived in [25] and used in the first experiments by Clauser and Freedman [40] (and later, others) to test whether the inequality is violated in the quantum world – it is. This variant is called the *CHSH inequality*, after the four authors (Clauser, Horne, Shimony, and Holt) on [25].[25]

[25] In 2022, John Clauser shared the Nobel prize in physics for this work. This long-delayed recognition is a Cinderella story, where Clauser's under-the-radar work (with equipment partly obtained by "dumpster diving" [11]) was initially denigrated by many physicists (other

The derivation of the CHSH inequality found in textbooks (e.g., [79]) usually involves the concept of *expected value* and a fundamental theorem in probability theory called *the linearity of expectation*, although the expositions don't always make that point explicitly. Its use makes the derivations shorter, but, I promised that in this book the only required mathematical background is arithmetic and simple logic, and so we will derive the CHSH inequality from scratch, without requiring any knowledge of expected values or the linearity of expectation. We do this through a series of exercises.

Suppose that $\overline{Q} = (q_1, q_2, ..., q_n)$ and $\overline{R} = (r_1, r_2, ..., r_n)$ are two ordered lists of n numbers each, where each number is either -1 or $+1$. Let q be the average of the n numbers in $\overline{Q}$, and let r be the average of the numbers in $\overline{R}$.

Then, let $\overline{Q} + \overline{R}$ denote the ordered list of sums $(q_1 + r_1, q_2 + r_2, ..., q_n + r_n)$, and let p be the average of the n sums in $\overline{Q} + \overline{R}$.

20a. Is it true that the average of the sums in $\overline{Q} + \overline{R}$ is the average of the numbers in $\overline{Q}$ plus the average of the numbers in $\overline{R}$, that is, that $p = q + r$? Hint: yes it is. Explain.

Now, let $\overline{Q} - \overline{R}$ denote the ordered list of n differences $(q_1 - r_1, q_2 - r_2, ..., q_n - r_n)$.

20b. Is it true that the average of the differences in $\overline{Q} - \overline{R}$ is the average of the numbers in $\overline{Q}$ minus the average of the numbers in $\overline{R}$, that is, $q - r$? Explain.

21. Next, suppose we have two additional ordered lists of n numbers each:

$\overline{S} = (s_1, s_2, ..., s_n)$ and $\overline{T} = (t_1, t_2, ..., t_n)$, where again, each number is either -1 or $+1$.

We use $\overline{QS}$ to denote the ordered list of n numbers $(q_1 \times s_1, q_2 \times s_2, ..., q_n \times s_n)$; and we use $\overline{QS}_i$ to denote the ith term in the list $\overline{QS}$, that is, the term $q_i \times s_i$. Similarly, we use $\overline{RT}$ to denote the ordered list of numbers $(r_1 \times t_1, r_2 \times t_2, ..., r_n \times t_n)$; and we use $\overline{RT}_i$ to denote the ith term in the list $\overline{RT}$, that is, the term $r_i \times t_i$.

Finally, we use $\overline{QS} + \overline{RT}$ to denote the ordered list of n numbers $(\overline{QS}_1 + \overline{RT}_1, \overline{QS}_2 + \overline{RT}_2, \ldots, \overline{QS}_n + \overline{RT}_n)$.

Is it true that the average of the n numbers in $\overline{QS} + \overline{RT}$ is equal to the average of the numbers in $\overline{QS}$ plus the average of the numbers in $\overline{RT}$? Explain.

than Bell and a few others) as unserious and unnecessary. Richard Feynman threw Clauser out of his office, "very offended that I should even be considering the possibility that quantum mechanics might not be giving the correct predictions" [24]. Clauser could not get an academic job in physics after doing these experiments. Some expositions on experiments testing Bell's inequalities (e.g., [102]) ignore Clauser entirely. For a more complete account of this history, see for example [11].

22. Similar to the definitions of $\overline{QS}$ and $\overline{RT}$, we define $\overline{RS}$ and $\overline{QT}$ as the ordered lists of n numbers $(r_1 \times s_1, r_2 \times s_2, ..., r_n \times s_n)$ and $(q_1 \times t_1, q_2 \times t_2, ..., q_n \times t_n)$, respectively. Recall that the possible values of each q_i, r_i, s_i, and t_i are only $+1$ and -1, for any i from 1 to n. Consider the sum $(q_i + r_i)$ and the difference $(r_i - q_i)$, for any i from 1 to n.

22a. Explain why the only possible values for these two terms are 0, -2, and +2. Further, explain why for each i exactly one of these two terms must have the value zero.

22b. Now explain that for each i from 1 to n, $(q_i + r_i) \times s_i + (r_i - q_i) \times t_i$ must be -2, 0, or $+2$, so $(q_i + r_i) \times s_i + (r_i - q_i) \times t_i$ must be at most $+2$. Hence: $(q_i \times s_i + r_i \times s_i) + (r_i \times t_i - q_i \times t_i) \leq 2$.

Define $\overline{QS} + \overline{RS} + \overline{RT} - \overline{QT}$ to be the ordered list of n numbers

$$(\overline{QS}_1 + \overline{RS}_1 + \overline{RT}_1 - \overline{QT}_1,\ \overline{QS}_2 + \overline{RS}_2 + \overline{RT}_2 - \overline{QT}_2,\ ...,\ \overline{QS}_n + \overline{RS}_n + \overline{RT}_n - \overline{QT}_n).$$

22c. Show that the average of the n numbers in $\overline{QS} + \overline{RS} + \overline{RT} - \overline{QT}$ is at most 2.

23. Let $A(\overline{QS})$, $A(\overline{RS})$, $A(\overline{RT})$, and $A(\overline{QT})$ denote the averages of the n numbers in each of the ordered lists $\overline{QS}$, $\overline{RS}$, $\overline{RT}$, and $\overline{QT}$, respectively.

Explain that

$$A(\overline{QS}) + A(\overline{RS}) + A(\overline{RT}) - A(\overline{QT}) \leq 2. \tag{3.14}$$

The inequality in (3.14) is the *CHSH inequality*. Explain why the CHSH inequality is true for any four ordered lists of n numbers each, where each number is either -1 or $+1$. Note that the CHSH inequality has been derived without any reference to quantum mechanics.

24. **The CHSH inequality and the EPR experiment of Clauser et al.** Now we relate the CHSH inequality to the EPR experiment described in [25, 40] and to the index-card model.

The experiment described in [25] and [40], which we will call a *CHSH-EPR experiment*, is a variant of the generic EPR experiment described in Section 2.2.1.1. In this variant, each detector can be oriented in only *two* ways instead of three. We use Q and R to denote the two ways that the first detector can be oriented, and S and T to denote the two ways that the second detector can be oriented. As before, in each trial, a detector randomly (with probability one-half) chooses one of its two possible orientations. When a particle hits a detector, the detector outputs either -1 or $+1$.[26]

An experiment consists of n trials. In the index-card model, when a particle is shot toward the first detector, it is given two rules, a Q-rule and an R-rule, telling the particle what number (-1 or $+1$) to cause the detector to output, depending on whether the first detector is in the Q or the R orientation. For example, if in the fifth trial, say, the index-card rule is ("$q_5 = +1, r_5 = -1$"), then, the particle should cause the first detector to output $+1$ if the detector is in the Q orientation and cause it to output -1 if the detector is in the R orientation.

Then, under the index-card model, in a CHSH-EPR experiment consisting of n trials, the lists $\overline{Q}, \overline{R}, \overline{S}$, and $\overline{T}$ record the $-1, +1$ values of the index-card rules given to the two particles in the n trials.

Explain that the CHSH inequality *must* hold for the four lists of values given to the index-card rules in a CHSH-EPR experiment.

25. **Determining if the CHSH inequality is violated in a real experiment** We derived the CHSH inequality above, which applies to the index-card model of a CHSH-EPR experiment. So, *if* an index-card theory agrees with the outcomes in a real CHSH-EPR experiment, then the outcomes should "conform" to the CHSH inequality. But what does that actually mean, since in any trial in a real experiment we only observe the results from *one* orientation of the first detector, either in the Q or R orientation, and from one orientation of the second detector, either in the S or T orientation.

Define q'_i to be the output (either -1 or $+1$) from the first detector *if* it is in orientation Q in the ith trial and r'_i to be the output *if* it is in orientation R in the ith trial. Similar definitions apply to s'_i and t'_i.

Then, we define $A'(\overline{QS})$ to be the average of the product of the numbers output by the two detectors, over those (generally, less than n) trials where the first detector is in orientation Q and the second detector is in orientation S. We similarly define $A'(\overline{RS}), A'(\overline{RT})$, and $A'(\overline{QT})$.

If a real, and long, CHSH-EPR experiment is guided by the index-card model, should we expect that

$$A'(\overline{QS}) + A'(\overline{RS}) + A'(\overline{RT}) - A'(\overline{QT}) \leq 2?$$

That is, if the CHSH-EPR experiment agrees with an index-card theory, is

$$A'(\overline{QS}) + A'(\overline{RS}) + A'(\overline{RT}) - A'(\overline{QT})$$

[26] As in other discussed experiments, we do not need to describe the inner workings of the detectors, nor the angles of the specific orientations Q, R, S, and T.

a good approximation for

$$A(\overline{QS}) + A(\overline{RS}) + A(\overline{RT}) - A(\overline{QT})$$

in a long experiment? Explain.

Hint: Remember that in each trial, each detector chooses one of the two possible orientations with equal probability, and under the index-card model, the choices of the two detectors are independent. To answer the question, review what was discussed and learned in Exercises 3 through 10.

The rest of the story Real CHSH-EPR experiments have shown violations of the CHSH inequality. With the appropriate choice of orientations Q, R, S, and T, $A'(\overline{QS}) + A'(\overline{RS}) + A'(\overline{RT}) - A'(\overline{QT})$ has been observed to be as large as $2\sqrt{2}$, which is what quantum mechanics predicts, in violation of the CHSH inequality.

4
Arrow's (and Friends') Impossibility Theorems

Of all the impossibility theorems discussed in this book, probably the one with greatest "name recognition" is *Arrow's impossibility theorem for fair voting* or more generally for *social choice*. Although many people have heard of this theorem, I expect that few can actually state it, and many fewer have ever studied any proof of it.

Arrow's theorem is central in Voting Theory, Economics, Political Science, Social Welfare, and so forth. Quoting from Sen [96]:

> Virtually all models to analyse political competition originate in some way or the other from Arrow's impossibility result.

So, Arrow's impossibility theorem has high impact. But for me, one of the most amazing things about Arrow's theorem (and this is true for *all* of the impossibility theorems fully proved in this book) is that its proofs only involve basic, step-by-step, logical reasoning (although a fair amount of it) – they do not require any advanced logic or mathematics.

4.1 Are Fair Elections Possible?

The most concrete framework for discussing Arrow's theorem, and its relatives, is in the context of election fairness, and that is the framework that we adopt here.

Ranked-choice voting: The election model Each of n voters provides an *ordered list* stating their ranking of m candidates, no ties allowed. Such a list is called a *preference list*. A collection of the preference lists from all the n voters is called a *profile*, denoted Γ. See Figure 4.1 for an example of a profile.

Profile Γ	Voters preference lists		
	1	2	3
	D	A	D
	A	C	C
	B	B	B
	C	D	A

Figure 4.1 A profile Γ with three voters $\{1, 2, 3\}$ and four candidates $\{A, B, C, D\}$.

Review question: Suppose we add a fourth voter to the example in Figure 4.1. Create a preference list for this new voter that is different from each of the three existing lists in Γ.

One such new list is (C, D, A, B). This list also has the extreme property that at every position in the new list, the candidate differs from the candidate at that position in each of the three existing lists. Verify this claim.

Can you find another suggested list for the fourth voter that also has this extreme property? If not, why not?

Election mechanisms To have an election, there must be a specified "election mechanism" (also called a *method* or *algorithm* or *scheme*) that takes in *any* possible profile and produces an *election outcome*, which (in Arrow's model) is an *ordered ranking* of the m candidates, with no ties. We will abstractly refer to an election mechanism as *EM*.

It is intuitive that different election mechanisms can lead to different election outcomes, so the choice of an election mechanism is extremely important. And how a mechanism is implemented and by whom are also critical. As Joseph Stalin said in a cynical, menacing quote:

> Those who cast the votes decide nothing. Those who count the votes decide everything.

Review question: Consider the following election mechanism to determine the election outcome for a profile. Give four points to a candidate any time that candidate is the *first* choice of a voter, three points to a candidate any time they are the *second* choice of a voter, two points when they are the *third* choice of a voter, and one point every time a candidate is the *fourth* choice of a voter. Then, for each candidate, add up the total number of points given to them.

Applying this election mechanism to profile Γ shown in Figure 4.1, how many points does each candidate get? Are there any ties? If the election mechanism orders the candidates based on the number of points each receives (largest first), what is the election outcome for Γ?

Condercet election mechanism Early in the study of ranked-order elections, examples were constructed that show the difficulty of designing sensible election mechanisms. Consider the following proposed election mechanism in an election with an *odd* number of voters:

> For each pair of candidates, (X, Y), require that X be placed above Y in the election outcome if more than half the voters prefer X to Y; otherwise require that Y be placed above X in the outcome. If there is an election outcome that obeys each of these requirements, it is called a *Condercet ranking.*

Verify that the Condercet ranking for profile Γ in Figure 4.1 is (D, A, C, B).

Now consider the profile shown in Figure 4.2. Is there a Condercet ranking for this profile? If not, why not?

So, it seems that the proposed Condercet mechanism has an undesirable property: It will not always give a Condercet ranking and hence will not always produce an election outcome. But, it is only one of many possible election mechanisms. Arrow's theorem concerns *all* possible mechanisms.

Is a fair election mechanism possible for ranked-choice voting? The issue that Arrow investigated is whether it is possible to have an election mechanism that obeys certain requirements (discussed below) that seem desirable for "fair" ranked-choice elections. Arrow proved that no such election mechanism can exist – it is *impossible* – as long as there are at least three candidates (and any number of voters). That is *Arrow's Impossibility Theorem.*

Arrow's theorem helped to start an entire field called "Social Choice Theory". Additional impossibility theorems for other types of voting (we discuss

Voters preference lists		
1	2	3
A	B	C
B	C	A
C	A	B

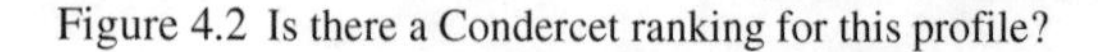

Figure 4.2 Is there a Condercet ranking for this profile?

one below) have been derived since Arrow's seminal 1950 paper. In 1972, Kenneth Arrow shared the Nobel prize in Economics for his impossibility theorem and for his role in founding Social Choice Theory.

4.2 First, a Related Impossibility Theorem

Before discussing the details of Arrow's impossibility theorem and proof, I want to discuss a related impossibility theorem, the Gibbard–Satterthwaite (GS) theorem, based on a voting model that I find more natural than Arrow's model, and which has a somewhat simpler proof. The backgrounds for the two theorems are related, and the styles of the proofs are similar. But each proof is self-contained, so you can skip either of these proofs and still be able to understand the other.[1]

The GS impossibility theorem concerns *ranked-choice* elections where each voter provides a rank ordering of the candidates, as in Arrow's model, but where the election mechanism declares a *single winner*, rather than declaring a rank-ordered *list* of the candidates, as in Arrow's model. We call this the *single-winner* model.

4.2.1 Requirements for a Fair Single-Winner Election Mechanism

The GS model of a fair single-winner election mechanism requires that the election mechanism obey three requirements (axioms). The first two requirements are *Unanimity* and *No-dictatorship*, defined below. The actual third GS requirement will be introduced in a bit, but for now, we will use a different third requirement called *Consistency*.

Three Requirements

1. **The Unanimity requirement:** For any profile, Γ, if there is a candidate, say X, who is at the top of *every* voter's preference list, then the election mechanism must declare candidate X the election winner.

[1] The GS theorem came about 25 years after Arrow's theorem, proved independently by Gibbard and by Satterthwaite in 1973 and 1975, respectively. Interestingly, the proof I will discuss for the GS theorem (due to P. Reny [88]) is based on a proof of Arrow's theorem due to John Geanakopolos. Reny showed how to convert Geanakopolos's proof of Arrow's theorem into a proof of the GS theorem.

The publication dates of these papers are confusing because Geanakopolos's paper was published in 2005, and yet Reny's paper, which is based on Geanakopolos's paper, was published in 2001. The resolution of this confusion is that an early version of Geanakopolos's paper was made public in 1996.

Profile Γ	Voters preference lists		
	1	2	3
	D	A	D
	A	C	C
	B	B	B
	C	D	A

Profile Γ'	Voters preference lists		
	1	2	3
	D	A	D
	B	C	A
	A	D	C
	C	B	B

Figure 4.3 Profile Γ, repeated from Figure 4.1, and the modified Profile Γ'. Suppose that the election mechanism *EM* declares candidate *D* to be the election winner for profile Γ. In every voter's list in profile Γ', candidate *D* is above every candidate that *D* is above in Γ. Note that compared to profile Γ, the orders of the candidates below *D* in Γ' have changed in the lists of voters 1 and 3. Also, in the list of voter 2, candidate *B* is now below *D* in Γ', but was above *D* in Γ. Since *EM* declared *D* the winner for profile Γ, consistency requires that *EM* also declare *D* the winner for profile Γ'.

2. **The No-dictator requirement:** A voter *i* is called a *dictator* if for *every* profile Γ, the election mechanism simply declares the winner of the election to be the top candidate in the list of voter *i*. The *No-dictator* requirement says that a fair election mechanism must *not* allow any voter to be a dictator.
3. **Consistency**[2]
   Given a profile Γ, and an election mechanism denoted *EM*, suppose *EM* declares a candidate *X* the winner for Γ. Consider another profile Γ' where in *each* voter's list, if candidate *X* is above a candidate *Y* in profile Γ, then *X* is above *Y* in profile Γ'. That is, suppose *X* is the election winner for profile Γ, and that in every voter's list for Γ', *no* candidate below *X* in Γ moves to be above *X* in Γ'.[3] When these conditions hold, we say that Γ and Γ' are *consistent for X*.

The *Consistency requirement* for *EM* says that if *EM* declares a candidate *X* the winner for profile Γ, and Γ and Γ' are consistent for *X*, then *EM* *must* also declare *X* the winner for profile Γ'. See Figure 4.3.

[2] The term "consistency" is more often called "monotonicity" in the voting literature, but I think "consistency" is a more understandable term.
[3] But, in any voter's list, some candidates above *X* in Γ might move below *X* in Γ'. Also, in any voter's list, the *order* of the candidates below (and also above) *X* may be different in Γ and Γ'.

Review question: Suppose that *EM* declares candidate *A* the winner for profile Γ shown in Figure 4.1. Would the consistency requirement be violated if *EM* does not declare candidate *A* the winner for profile Γ' shown in Figure 4.3? What if *EM* declares *B* or *C* the winner for Γ?

Is Consistency Sensible? Consistency seems like a sensible requirement for a fair election mechanism. After all, consider *in*consistency: Suppose a candidate X wins for a profile Γ, where in the list of each voter i, X is above a set of candidates denoted C_i; and then in Γ', X is again above all candidates in set C_i (and maybe even more) in the list of each voter i, and yet, the election mechanism does *not* declare X the winner for Γ'. Surely X would then be outraged and bellow: "But, in each voter's list, I beat the same set of candidates in both Γ and Γ'. Why should I win in one profile, but lose in the other?"

Review question: What do you think of the above argument in favor of the consistency requirement? Try to construct two profiles Γ and Γ', which together suggest that a fair election mechanism should *not* always obey the consistency requirement.

We will now prove:

Theorem 4.2.1 *There is no* single-winner *election mechanism that simultaneously obeys the Unanimity, No-dictator, and Consistency requirements when there are at least three candidates. Such an election mechanism is* impossible.

In order to prove Theorem 4.2.1, we first need a definition and a lemma. Consider the two profiles, Γ and Γ' shown in Figure 4.4. Notice that in the lists of voter 1, candidate X is *immediately* above candidate Y in Γ, but Y is immediately above X in Γ'. Otherwise Γ and Γ' are identical. This example leads to the following definition:

(X, Y)-reversal Suppose there are two profiles Γ and Γ', and two candidates, X and Y, where X is *immediately* above Y in the preference list of some voter i in profile Γ, and Γ' is created from Γ by exchanging X and Y in the list of voter i. Otherwise, Γ and Γ' are identical. Then, we say that Γ' is an *(X, Y)-reversal* of Γ. Symmetrically, Γ is an (Y, X) reversal of Γ'.

Γ			Γ'		
1	2	3	1	2	3
X	W	Y	**Y**	W	Y
Y	X	W	**X**	X	W
Z	Z	X	Z	Z	X
W	Y	Z	W	Y	Z
	a			b	

Figure 4.4 In Γ, candidate X is immediately above Y in the list of voter 1; and Γ' is created by exchanging X and Y in the list of voter 1. Γ' is called an (X, Y)-*reversal* of Γ.

Lemma 4.2.1 *Suppose that the single-winner election mechanism EM declares candidate X to be the winner for a profile* Γ*; and that profile* Γ' *is an* (X, Y)*-reversal of* Γ. *If the election mechanism EM obeys the* Consistency *requirement, then EM must* either *declare X the winner* or *declare Y the winner for profile* Γ' *– no other winner is possible for* Γ'.

Proof For a proof by contradiction, suppose *EM* declares some other candidate, neither X nor Y, say Z, to be the winner for profile Γ'. Then, if we exchange the positions of X and Y in Γ', we will re-create profile Γ. In that switch of X and Y, the set of candidates below Z, in each voter's list, is exactly the same in Γ as in Γ'. So, since *EM* is assumed to obey the Consistency requirement, *EM* should declare candidate Z to be the winner for the re-created profile Γ.

But this is a contradiction because the original Γ is identical to the re-created Γ, and *EM* declared $X \neq Z$ to be the winner for the original Γ. Hence, Z cannot be the winner for Γ', and only X or Y can be the winner for Γ'. See Figure 4.5. ■

Review question: You might think that the proof of Lemma 4.2.1 also rules out Y as the winner for Γ', showing that only X could be the winner. That is not true. Explain why.

As a hint, try the following: Assume that *EM* declares Y to be the winner of the election for profile Γ', and try to reach the contradiction that Y must be the winner for Γ, as we did when we assumed that Z was the winner for Γ'. There is a step where that attempted proof will fail if Y is the assumed winner for Γ', but it did not fail when using Z as the assumed winner. Where exactly is that step? Explain why it fails if Y is the assumed winner.

Γ			Γ'			The re-created Γ		
1	2	3	1	2	3	1	2	3
X	W	Y	**Y**	W	Y	X	W	Y
Y	X	W	**X**	X	W	Y	X	W
Z	Z	X	Z	Z	X	Z	Z	X
W	Y	Z	W	Y	Z	W	Y	Z

Figure 4.5 Suppose the single-winner election mechanism *EM* (which we have not specified) declares *X* the winner for profile Γ. Suppose also that Γ' is an (*X*, *Y*)-reversal of Γ, and that *EM* declares some candidate, say *Z* for concreteness, other than *X* or *Y*, to be the winner for Γ'. Then, if *X* and *Y* again change places, re-creating profile Γ, the Consistency requirement (which we have assumed is obeyed by *EM*) says that *Z* should be the winner for Γ. But we know that *X* is the winner for Γ. Hence, *Z* can't be the winner for Γ', so only *X* or *Y* could be.

4.2.2 Now Back to Theorem 4.2.1

The game plan The approach to proving Theorem 4.2.1 is to show that *if EM* is any election mechanism that obeys the *Unanimity* and *Consistency* requirements, *then* there must be a dictator, violating the *No-dictator* requirement for *EM*. So, no election mechanism can obey all three of the GS requirements.

In more detail, we will create a *series* of profiles, starting at Γ_0, and for each profile we will deduce who must be, or could be, the winner for that profile. We will ultimately reach a profile Γ_4 where there is some voter called the *pivotal voter* and some candidate *A* such that *A* is at the *top* of the pivotal voter's preference list but at the *bottom* of all other preference lists, and yet, *EM* will declare candidate *A* the winner for profile Γ_4.

Declaring *A* the winner for profile Γ_4 seems pretty extreme and gives some evidence that the pivotal voter is a dictator, but it doesn't, by itself, prove it. However, we will repeat the series of profiles, with other candidates in place of *A*, and see that whichever candidate is at the top of the pivotal voter's list will be declared the winner by *EM*, no matter what the other details of the profile may be. That will establish the pivotal voter as a real dictator. That is the game plan.

4.2.2.1 Proof of Theorem 4.2.1

The starting profile in the series To start, let Γ_0 be a profile with candidate *A* on the top and candidate *B* at the bottom of every voter's preference list. All other details of Γ_0 are arbitrary. By the Unanimity requirement, *EM* must declare candidate *A* the winner. See the left panel of Figure 4.7.

Now modify profile Γ_0 by moving candidate *B* up to just below candidate *A* in the list of voter 1. By Consistency, candidate *A* will still be declared the

List of voter 1 in Γ_0	After the first move of B	After exchange of A and B
A	A	B
X	B	A
Z	X	X
Y	Z	Z
B	Y	Y

Figure 4.6 An example of the list of voter 1 in Profile Γ_0 and the two moves of candidate *B* in the list of voter 1. In Γ_0, candidate *A* is at the top of the list of voter 1 and candidate *B* is at the bottom. The positions of the other candidates are arbitrary and only specified to make a concrete example.

Γ_0				Γ_1				Γ_2			
1	2	3	4	1	2	3	4	1	2	**3**	4
A	A	A	A	**B**	**B**	A	A	B	B	**B**	A
X	X	Y	Y	A	A	**B**	Y	A	A	**A**	Y
Z	Z	Z	Z	X	X	Y	Z	X	X	Y	Z
Y	Y	X	X	Z	Z	Z	X	Z	Z	Z	X
B	B	B	B	Y	Y	X	B	Y	Y	X	B

Figure 4.7 Example of successive profiles Γ_0, Γ_1, and Γ_2. The characters shown in bold highlight the described changes. The pivotal voter in this illustration is voter 3. Since we have not fully specified the election mechanism, *EM*, this example is just to illustrate the process of creating Γ_2 from Γ_0 – the pivotal voter can't be deduced without fully specifying an election mechanism. Profile Γ_2 is the first profile where candidate *B* is declared the winner, and Γ_1 is the profile just before the positions of *A* and *B* were exchanged in the list of the pivotal voter.

winner by *EM*. Then, exchange the positions of candidates *A* and *B* in the list of voter 1. By Lemma 4.2.1, the winner declared by *EM* for this modified profile must either be candidate *A* or candidate *B*. See Figure 4.6.

If, after these moves in the list of voter 1, *EM* again declares candidate *A* the winner, repeat the same moves of *B* and *A*, but in the list of voter 2. As in the list of voter 1, after moving *B* to just below *A* in the list of voter 2, the Consistency requirement dictates that *A* will still be the winner. Then, after exchanging the positions of *A* and *B* in the list of voter 2, Lemma 4.2.1 dictates that only *A* or *B* can be declared the winner. If the winner is again candidate *A*, repeat the same moves of *B* and *A* in the list of voter 3. Continue such *B* and *A* moves, in successive voter's lists, until the *first* point where *EM* declares candidate *B* to be the winner. There must be such a point, for if the successive moves continue until *B* is the top of *every* voter's list, then by Unanimity, *EM* will declare *B* the winner.

Defining the pivotal voter At the first point that B is declared the winner, stop the moves; label the voter whose list was just modified the *pivotal voter*; and label the profile at that point Γ_2. Profile Γ_1 is the profile just *before* the positions of A and B were exchanged in the list of the pivotal voter. By the definition of Γ_2, candidate A is still the winner for profile Γ_1. See Figure 4.7.

Now we modify Γ_2 Now that we have identified the pivotal voter, we will modify profile Γ_2 and use the modification to help establish some facts about mechanism *EM* (which, recall, is assumed to obey the Unanimity and Consistency requirements).

From profile Γ_2, move candidate A *down* to the *bottom* of the preference list of every voter i to the *left* of the pivotal voter, and move candidate A *down* to just *above* candidate B in the list of every voter to the *right* of the pivotal voter. The list of the pivotal voter is unchanged. Call the resulting profile Γ_{2a}. See Figure 4.8.

Note that in the preference list of every voter in Γ_{2a}, candidate B is above every candidate that B is above in profile Γ_2. So, since B is the winner for profile Γ_2, the Consistency requirement dictates the following:

Fact 0: Candidate B must also be the winner in profile Γ_{2a}.

We will put this fact to the side for a bit, but use it shortly in our analysis of a profile derived from Γ_1.

Now we similarly modify Γ_1 We also modify profile Γ_1 in the same way we modified Γ_2, that is, by moving A to the bottom of each list to the left of the pivotal voter, and moving A to just above B in the list of every voter to the right of the pivotal voter, and leaving the list of the pivotal voter unchanged. Call the resulting profile Γ_{1a}. See Figure 4.9.

Γ_{2a}			
1	2	**3**	4
B	B	B	Y
X	X	A	Z
Z	Z	Y	X
Y	Y	Z	**A**
A	**A**	X	B

Figure 4.8 Profile Γ_{2a} derived from profile Γ_2. Recall that voter 3 is the assumed pivotal voter and that B is the winner for Γ_2. By Consistency, candidate B will still be the winner for profile Γ_{2a}.

	Γ_{1a}		
1	2	**3**	4
B	B	A	Y
X	X	B	Z
Z	Z	Y	X
Y	Y	Z	**A**
A	**A**	X	B

Figure 4.9 Profile Γ_{1a} derived from Γ_1. We will show that voter *A* is the winner for profile Γ_{1a}.

Claim: *EM* must declare candidate *A* the winner for profile Γ_{1a}.

To prove this claim, note first that the only difference between profiles Γ_{1a} and Γ_{2a} is that in the list of the pivotal voter, candidate *B* has moved from just below *A* in Γ_{1a} to just above *A* in Γ_{2a}. Then, since *B* is the winner for profile Γ_{2a} (as stated in Fact 0), if we exchange the positions of *A* and *B* in Γ_{2a} to re-create profile Γ_{1a}, Lemma 4.2.1 restricts the winner for profile Γ_{1a} to be either candidate *B* or candidate *A*. We will see that it cannot be *B*.

As observed earlier, in each voter's list, the set of candidates below *B* in Γ_1 is the same as in Γ_{1a}. So, by the Consistency requirement, if the winner for Γ_{1a} is *B*, then *B* must also be the winner for Γ_1. But that would be a contradiction, since we know that candidate *A* is the winner for profile Γ_1. So *B* cannot be the winner for profile Γ_{1a}, and *A* must be the winner, as claimed.

Recapping: The election mechanism *EM* will declare candidate *A* the winner for profile Γ_{1a} and will declare candidate *B* the winner for profile Γ_{2a}.

Now modify profile Γ_{1a} again Arbitrarily, pick a candidate, say *X*, other than *A* or *B*. Then, starting from profile Γ_{1a}, move (if necessary) candidates *A*, *B*, *X* so that:

1. For every voter *i* to the *left* of the pivotal voter, the *bottom* three candidates in the list of voter *i* are *X*, *B*, *A* in that order.
2. The *top* three candidates in the list of the pivotal voter are *A*, *X*, *B*, in that order.
3. For every voter *i* to the *right* of the pivotal voter, the *bottom* three candidates in the list of voter *i* are *X*, *A*, *B* in that order.

The resulting profile is called Γ_3. See Figure 4.10.

Claim: Candidate *A* is the winner for profile Γ_3.

The reason is that starting from profile Γ_{1a} (where *A* is the winner), in the creation of Γ_3 no candidate moves from below *A* to above *A* in any voter's list.

Γ_3 1	2	**3**	4
Z	Z	**A**	Y
Y	Y	**X**	Z
X	**X**	B	**X**
B	**B**	Y	**A**
A	**A**	Z	**B**

Figure 4.10 Example of profile Γ_3 created from Γ_{1a}, as described above.

Γ_4 1	2	**3**	4
Z	Z	A	Y
Y	Y	X	Z
X	X	B	X
B	B	Y	**B**
A	A	Z	**A**

Figure 4.11 Profile Γ_4 created from Γ_3, as described above.

In more detail, candidate *A* is at the bottom of the list in both Γ_{1a} and Γ_3, for every voter *i* to the *left* of the pivotal voter; and *A* is at the top of the pivotal voter's list in both Γ_{1a} and Γ_3; and candidates *A* and *B* are in the bottom two positions, in that order, for every voter to the *right* of the pivotal voter, in both Γ_{1a} and Γ_3. So, in every voter's list, candidate *A* is above the same set of candidates in both Γ_{1a} and Γ_3. Therefore, since candidate *A* is the winner for Γ_{1a}, the Consistency requirement dictates that candidate *A* must also be the winner for Γ_3.

Onward from profile Γ_3 Next, from profile Γ_3, for each voter to the *right* of the pivotal voter, exchange the positions of candidates *A* and *B*. The resulting profile is called Γ_4. Note that in Γ_4, candidate *A* is at the *bottom* of every voter's list, except for the pivotal voter where candidate *A* is at the *top*. See Figure 4.11.

We will show that despite being at the bottom of all but one list, *EM* will still declare candidate *A* the winner for profile Γ_4, as advertised in the "Game Plan" before the start of the proof.

Explaining why candidate *A* **is the winner for** Γ_4 Profile Γ_4 is created from Γ_3 (where *A* is the winner, and the bottom two candidates to the right of the pivotal voter are *A* and *B* in that order) by exchanging the positions of candidates *A* and *B* in every list to the right of the pivotal voter. Now suppose we make this transformation from Γ_3 to Γ_4 by doing the exchanges in the lists of one voter at a time.

Starting from Γ_3, let voter i be the first voter to right of the pivotal voter, and exchange the positions of candidates A and B in the list of voter i. Since A is the winner for Γ_3, this modified profile is an (A, B)-*reversal* of Γ_3. Therefore, Lemma 4.2.1 dictates that after this exchange, either candidate A or B will be the winner. Whichever of these two candidates is the winner, when we next exchange the positions of A and B in the list of voter $i + 1$, Lemma 4.2.1 dictates that *EM* must again declare either A or B the winner. Continuing in this way, successively switching A and B in the lists of the voters to the right of the pivotal voter, and applying Lemma 4.2.1 each time, we see that the winner for Γ_4 must either be candidate A or B. We will next argue that it *can't* be candidate B.

To B or not to B? Not to B! Note that candidate X is above B for every voter in profile Γ_4. So, from Γ_4, if we raise X to the top of each voter's list, the set of candidates below B would remain the same. Therefore, *if EM* declares candidate B the winner for Γ_4, the Consistency requirement would dictate that B must continue to be the winner after X is raised to the top of each voter's list. But, if X is the top of each voter's list, then the Unanimity requirement would dictate that candidate X would be the winner at that point. This is a contradiction.

The only way out of this contradiction is to conclude that *EM* does *not* declare candidate B the winner for Γ_4, and since we already established that the winner must be either A or B, candidate A must be the winner for Γ_4, as claimed.

Ready for a punchline? In profile Γ_4, candidate A is at the top of the list of the pivotal voter but at the *bottom* of the list of every other voter. And yet, despite so much losing, *EM* declares candidate A the winner for profile Γ_4.

Further, we can *permute*, in any way, the orders of all non-A candidates in all the voters' lists, and candidate A would still be at the top of the list of the pivotal voter, and at the bottom of all other lists. So, in each list, the set of voters below A would always be the same as in Γ_4. Hence, by the Consistency requirement, candidate A would be the winner in every one of those permuted profiles.

Moreover, in any of those permuted profiles, if we move A up from the bottom of any list, the Consistency requirement would again dictate that candidate A will be the winner. The only thing that is constant over all of those profiles is that A is at the top of the list of the pivotal voter. And so we have the punchline:

> In any profile (with at least three candidates) where candidate A is at the top of the preference list of the pivotal voter (and all other details of the profile are arbitrary), any voting mechanism *EM* that obeys the Unanimity and Consistency requirements will declare A the election winner.

So, the pivotal voter is, at least, a *mini-dictator* (for candidate A). We will next argue that the pivotal voter is not just a mini-dictator but a real dictator.

Almost done. Rinse, rename, and repeat What's in a name? Candidate A is not really special![4]

Consider the above argument (starting from Γ_0 and leading to Γ_4), establishing that the pivotal voter is a mini-dictator for candidate A. But now, literally change every occurrence of the character A to B and every occurrence of character B to A, in that argument. The result will be a correct argument that exactly the *same* voter as before will be found to be the pivotal voter and a correct argument that the pivotal voter is a mini-dictator for candidate B. That is, whenever B is at the top of the preference list of the pivotal voter, the election mechanism *EM* will declare B the election winner. This is called a *renaming* argument.

OK, but we can repeat the renaming argument with any candidate (other than A or B) playing the part of A. We use C to denote that candidate. Then, following the original argument (that the pivotal voter is a mini-dictator for candidate A), literally change every occurrence of character A to C and every occurrence of C to A in the argument. The result will again be a correct argument that the previous pivotal voter (for A and B) will again be the pivotal voter for C and that the pivotal voter is a mini-dictator for candidate C.

So, the pivotal voter is a mini-dictator for *every* candidate, and that makes the pivotal voter a real dictator. That is, we have proved:

> Assuming that the *EM* obeys the Unanimity and Consistency requirements, for *any* profile with at least three candidates, *EM* will declare the election winner to be whichever candidate is at the top of the preference list of the pivotal voter. None of the other details of the profile matter. The pivotal voter must be a dictator.

And this concludes the proof of Theorem 4.2.1. ■

Hooray! We have proved Theorem 4.2.1 and have only used careful, step-by-step reasoning – no fancy logic and no advanced math. The argument was

[4] We are at a point in the proof of Theorem 4.2.1 where a mathematician writing for a mathematically "mature" audience might just say: "And the Theorem follows by renaming." But, I want to make this renaming idea explicit and help all readers develop "mathematical maturity".

long, and you had to read it actively, verifying each step for yourself, but (I hope) each step was clear and relatively simple.

Review question The statement of Theorem 4.2.1 requires that there be at least three candidates. Therefore, the proof we gave of Theorem 4.2.1 must fail if we try to modify it to work with only two candidates.

Where exactly is that failure, and why exactly does the proof fail if we try to make it work for only two candidates?

Review question The statement of Theorem 4.2.1 does not specify a minimum number of voters, implying that the theorem is correct for any number of voters.

Explain why the theorem is true if there is only *one* voter.

Go through the proof of Theorem 4.2.1 to verify that it is correct when there are only *two* voters (and at least three candidates). The proof might simplify for that case. Find those simplifications.

4.2.3 Back to GS

Theorem 4.2.1 is close to the GS theorem, but it isn't the GS theorem. To state and prove the GS theorem, we need the concept of an election mechanism being *deceit-immune*. We introduce that concept with an example of an election mechanism that is *not* deceit-immune. Consider the following:

Strange election mechanism, *EM*: If every voter has the same candidate at the top of their list, then *EM* declares that candidate the winner. Otherwise, *EM* declares the winner to be the candidate who is in the *second* position in the list of voter 1.

Review question: Does this strange *EM* obey the *Unanimity* requirement? Does it also obey the *No-dictator* requirement? Explain.

Now suppose voter 1 knows how *EM* works and sees the submitted lists of all the other voters before voter 1 submits a list. And, suppose that candidate *X* is the *true* first choice of voter 1. How can voter 1 ensure that *EM* declares candidate *X* the winner?

Voter 1 should create a list where candidate *X* is in *second* position. Then, after seeing the submitted lists of the other voters, if all those submitted lists have the same first choice, say *Y*, then voter 1 should arbitrarily put some candidate other than *X* or *Y* in the first position of their list, so that there is no unanimous first choice. If the other submitted lists do not all have the same first choice, then voter 1 can arbitrarily put any candidate, other than X, in the first position. All other details of the list of voter 1 are arbitrary.

In this way, voter 1 is deceptive because they do not submit a list containing their true preferences, but instead they submit a list where at least the first two entries are dishonest – candidate X is not the true second choice of voter 1, but their first choice.

Voter 1 benefits from deceit If voter 1 creates and submits a list as described, then candidate X is *guaranteed* to be declared the winner. Candidate X *might* have been declared the winner if voter 1 had submitted their honest list (where X would have been the first choice), but with the described deceptive behavior, voter 1 will never do worse and can sometimes do better. In this case, we say that voter 1 *benefits* by being deceptive.

Review question: We could have added the following case to the above strategy of voter 1: If all the other voters put X at the top of their lists, then voter 1 should submit an honest preference list, which also puts X at the top of their preference list.

Would this modified strategy still ensure that *EM* always declares candidate X the winner? Why do you think we did not include this case in the strategy of voter 1?

4.2.3.1 Deceit-Immune Election Mechanisms

An election mechanism, *EM*, where *no voter* can *ever* benefit by being dishonest, even if they know exactly how *EM* operates, and always know the submitted preference lists of all the other voters before submitting a list of their own, is called *deceit-immune*.[5] With a deceit-immune election mechanism, no voter has any incentive to submit a dishonest preference list. Honesty, then, really is the best policy – or at least no worse than any other.

More formally, an election mechanism, *EM*, is *not* deceit-immune (in which case, *EM* is also called *manipulable*) if there exist two profiles Γ and Γ' which are identical except for the list of one voter, say voter i, such that:

1. *EM* declares a candidate X the winner for Γ' and declares a different candidate Y the winner for Γ.
2. In profile Γ' (be sure you read Γ' here, not Γ) candidate Y is above candidate X in the list of voter i.

[5] The term "strategy proof" is more often used, but I don't like it.

Review question Using the formal definition for what it is for *EM* to be manipulable, how does voter i benefit from being deceitful when their true preference list is the one in profile Γ?

Deceit-immunity is a good thing Having a deceit-immune election mechanism is very desirable because it simplifies the task of each voter, avoiding the games and double-think (or triple-think) that voters might otherwise engage in. When voters do engage in that kind of double-think, always suspicious of the other voters, they eventually become dissatisfied with the voting mechanism. So, it is standard in game theory and in the design of election and auction mechanisms to try to create deceit-immune mechanisms. There is a large and active literature in computer science and economics on those efforts. There have been some successes, but also some failures – the GS theorem explains one kind of failure.[6]

4.2.4 The Actual GS Impossibility Theorem

Theorem 4.2.2 *(GS) There is no single-winner election mechanism that is* deceit-immune *and obeys the* Unanimity *and the* No-dictator *requirements. Such an election mechanism is* impossible.

In order to prove Theorem 4.2.2, we first establish the following:

Lemma 4.2.2 *If a single-winner election mechanism is deceit-immune, then it must obey the Consistency requirement.*

Proof Consider a profile Γ which reflects the true preferences of the voters, where candidate A is declared the winner by *EM*; and let Γ' be a different profile where Γ and Γ' are consistent for candidate A. That is, for every voter i, any candidate below A in the preference list of voter i in Γ is also below A in the preference list of voter i in Γ'. Then, to show that *EM* obeys the Consistency *requirement*, we need to show that *EM* would again declare A the winner for profile Γ'.

[6] Election mechanisms where voters have an incentive to not submit their true preferences, are examples of systems that have a more general property called *incompatible incentives*: participants in the system can benefit by behaving in a way that looks to be contrary to their best interests. A beautiful, almost shocking, example of this occurred in the most recent (February 2023) Super-Bowl (American football), where the player with the ball stopped by himself a few yards short of the goal line, instead of crossing it to make a touchdown; and once that player was close to the goal line, the defensive players did not try to keep him from scoring – they wanted him to score. The rules of football gave incentives in that moment for such odd, reversed play [26].

L_1	L'_1
.	A
.	.
A	B
.	.
B	.
.	.

Figure 4.12 *EM* is assumed to be deceit-immune (i.e., not manipulable), and profiles Γ and Γ' are assumed to be consistent for candidate *A*, the declared winner for Γ. If *EM* declares $B \neq A$ the winner for Γ', then *A* must be above *B* in list L_1, or else the replacement of L_1 with L'_1 would result in a winner (i.e., *B*) where voter 1 benefits compared to the result (winner *A*) for Profile Γ. But, that would then establish that *EM* is manipulable, contrary to the assumption that *EM* is deceit-immune.

Proving that *A* will be declared the winner for Γ'. We start with the special case that Γ and Γ' differ only in the preference lists of a *single* voter, say voter 1. The two lists of voter 1 in Γ and Γ' are called L_1 and L'_1, respectively. All other details of Γ and Γ' are assumed to be identical in this special case. See Figure 4.12.

Now suppose that *EM* does *not* obey the Consistency requirement and does not declare candidate *A* the winner for profile Γ'. Instead, *EM* declares some other candidate, say *B*, the winner for Γ'. In that case, we claim that candidate *A must* be above *B* in list L_1. To see this, note that if voter 1 submits list L'_1 instead of L_1, the result would be that *EM* actually receives profile Γ' and will (as assumed above) declare candidate *B* the winner. So, if *B* is above *A* in L_1, profiles Γ and Γ' show (using the formal definition) that *EM is* manipulable, contradicting the assumption that *EM* is deceit-immune.

To review, assuming that *EM* is deceit-immune, if *EM* declares some candidate $B \neq A$ the winner in Γ', candidate *A* must be above *B* in list L_1. It then follows, since *A* is the winner for Γ, and Γ and Γ' are assumed to be consistent for candidate *A*, that *A* must also be above *B* in list L'_1. See Figure 4.12 for a summary of this situation.

We can now finish the argument that *A* will be the winner for Γ', by essentially reversing the roles of Γ and Γ' in the above argument.

Remember that we are still in the case that the winner for Γ' is assumed to be candidate $B \neq A$, and we just established that *A* is above *B* in list L_1. So, if voter 1 submits the list L_1 instead of L'_1, *EM* will actually receive profile

Γ, and since we know that *EM* declared candidate A the winner for Γ, it will again declare A the winner, instead of B. And, we have also established that A is above B in profile Γ', so (by the formal definition) *EM* would be manipulable, (voter 1 benefits by substituting Γ in place of his true list, Γ', contradicting the assumption that *EM* is deceit-immune).

Summarizing what we now know We assumed that *EM* is deceit-immune, and that profiles Γ and Γ' differ only in the preference list of voter 1. If *EM* declares a candidate, say A, the winner for Γ, and Γ and Γ' are consistent for A, then *EM* must also declare A the winner for profile Γ'. This supports the correctness of Lemma 4.2.2 but does not fully prove it. We do that next.

Generalizing To finish the proof of Lemma 4.2.2, we generalize the argument used in the above special case, to the full case that Γ' differs from Γ in *more* than one voter's list. Again, we assume that Γ and Γ' are consistent for candidate A, the winner for Γ, and that *EM* is deceit-immune.

We want to show that *EM* will also declare A the winner in Γ' to establish that *EM* obeys the Consistency requirement. This is actually simple: Starting with voter 1, we sequentially transform Γ to Γ' one voter list at a time. That is, in each step, we replace a list in Γ with the corresponding list in Γ'. After the list for a voter i has been replaced, the modified profile is called Γ_i.

We proved above that when we replace list L_1 in Γ with list L_1' from Γ', but leave the rest of Γ the same, *EM* must declare candidate A the winner of the modified profile, which is what we now call profile Γ_1. Well, we can next make the same change for voter 2, replacing, in Γ_1, the list L_2 (of Γ) with the list L_2' from Γ', creating the profile Γ_2. Since Γ_1 and Γ_2 differ in only one list, namely the list of voter 2, the same argument as above shows that *EM* must again declare A the winner for Γ_2. Continuing in this way, candidate A must be declared the winner for each Γ_i, so after Γ has been fully transformed to Γ', A will still be the winner. This means that *EM* obeys the Consistency *requirement*, proving Lemma 4.2.2. ■

Finally, Lemma 4.2.2 immediately leads to a proof of the GS theorem.

Proof of the GS Theorem 4.2.2: If *EM* obeys the *Unanimity* requirement and is *Deceit-immune*, then by Lemma 4.2.2, *EM* also obeys the *Consistency* requirement. So by Theorem 4.2.1, *EM* must allow a dictator, violating the *No-dictator* requirement and proving the GS Theorem 4.2.2. ■

4.2.5 A Small Sermon about "Useless" Knowledge

It is sometimes tempting to feel sure that some knowledge is useless or uninteresting. But, I believe those views are misguided, and we never know where knowledge will lead. We saw this in our discussion of the EPR paper, where Abraham Pais doubted that the content of that paper would ever have any impact on the future of physics. And, we saw it in the unexpected way that Bell's *impossibility* theorem and inequality are now used in experimental methods that exploit entanglement.

Another example of useful–useless knowledge is Theorem 4.2.1, involving the Unanimity, No-dictator, and Consistency requirements for a fair single-winner election mechanism. As discussed earlier, the Consistency requirement seems sensible in some contexts, but it is controversial because it may sometimes be undesirable. People who think that consistency is not desirable might then say that Theorem 4.2.1 is useless or uninteresting.

However, even if the Consistency requirement is not sensible, Theorem 4.2.1 is still valuable because it is at the core of our proof of Theorem 4.2.2, where the Consistency requirement has been replaced with the Deceit-immunity requirement. While there may be some types of elections where Deceit-immunity is not needed, there is no question that deceit-immunity is a widely desired and appropriate property of an election mechanism, and so the questions of relevance raised for Theorem 4.2.1 are not raised for Theorem 4.2.2. Theorem 4.2.1 is thus either useful for what it actually implies about election mechanisms requiring consistency, or as a technical tool to prove less controversial statements about election mechanisms requiring deceit-immunity, or both. We will return to this sermon with additional examples later.

4.3 Arrow's Impossibility Theorem

Personally, I think that the GS impossibility theorem is more interesting than Arrow's impossibility theorem, but it was Arrow's theorem that got everything started and that still gets most of the glory. For that reason, and because the modern proofs are delightful, we present Arrow's impossibility theorem, and a proof of it, in detail.

Recall the Arrow model In Arrow's election model, an election mechanism, EM, is given a profile Γ consisting of the preference lists of the voters, exactly as in the *single-winner* (GS) election model. But an election outcome consists of an *ordered* list of *all* the candidates, rather than just one winner. See again Figure 4.1.

We start with a definition We will often talk about the *relative order* of *two* candidates in a preference list, or in an election outcome. The relative order of two candidates specifies which candidate is ranked *above* the other (and hence which candidate is ranked *below* the other), rather than specifying their actual positions in a list. For example, if an ordered list of four candidates is (B, A, D, C), then the relative orders of the six pairs of candidates (A, B), (A, C), (A, D), (B, C), (B, D), and (C, D) are:

> B above A, A above C, A above D, B above C, B above D, and D above C.

Now an essential technical point that may not be obvious:

Fact 1: If you know the relative order of *every* pair of candidates in an ordered list, but you haven't been shown the list itself, then you can determine with certainty what that ordered list must be. A more mathematical way to state this is:

> The relative orders of all the pairs of candidates in an ordered list uniquely determine the order of the list.[7]

4.3.1 Arrow's Requirements for a Fair Election Mechanism

Arrow initially described five requirements that any "fair" ranked-choice election mechanism should obey. Over time, these have been reduced to just three: *Unanimity, No-dictator, and Independence of Irrelevant Alternatives (IIA)*. In the context of Arrow's theorem, the Unanimity and No-dictator requirements are *generalizations* of the requirements with the same names in the single-winner model, and the IIA requirement is related to the Consistency requirement. We now discuss these requirements in depth.

1. **Arrow's Unanimity requirement:** For any profile Γ, if *all* voters prefer a candidate X to a candidate Y in profile Γ, then a fair election mechanism *must* rank X above Y in its election outcome for profile Γ.

   In the rest of this chapter, the word *Unanimity* will refer to the definition given here, rather than to the definition used for single-winner elections.
2. **Arrow's No-dictator requirement:** In the context of Arrow's theorem, a voter i is called a *dictator* if for every profile Γ, the election mechanism

[7] One way to determine the order of an originally unknown list is to look at the relative order information (one relative order per pair of candidates) to find the (unique) candidate who is above all the other candidates. That candidate must be at the top candidate in the unknown list. Then, remove all of the pairs of candidates that contain that top candidate and find the candidate who is above all other candidates in the remaining pairs. That candidate is the second candidate in the unknown ordered list. Repeat these operations, finding the third candidate in the list, and so on until the list is fully ordered.

Profile Γ'	Voters preference lists		
	1	2	3
	B	A	C
	D	B	A
	A	C	B
	C	D	D

Figure 4.13 A profile Γ' with the same three voters and four candidates as the profile Γ in Figure 4.1. Comparing those profiles, we see that voters 1 and 2 prefer candidate *A* to candidate *C* in both profiles; and voter 2 prefers candidate *C* to candidate *A* in both profiles. The IIA requirement then says that the relative order of *A* and *C* must be the same in any election outcome produced for profiles Γ and Γ'.

produces an election outcome that is *identical* to the preference list of voter *i* in Γ.

The *Arrow's No-dictator requirement* says that a fair election mechanism must *not* allow any voter to be a dictator.

In the rest of this chapter, the word *dictator* will refer to the definition given here, rather than to the definition used for single-winner elections.

3. **The Independence of Irrelevant Alternative (IIA) requirement:** Suppose that for any two profiles Γ and Γ', and any two (arbitrary, but fixed) candidates, say *A* and *B*, *each* voter ranks *A* above *B* in Γ if and only if that voter ranks *A* above *B* in Γ'. That is, suppose for each voter *i*, the relative order of candidates *A* and *B* in the preference list of voter *i* in Γ is the *same* as the relative order of *A* and *B* in the preference list of voter *i* in Γ'.

   The IIA requirement says that when the above conditions (suppositions) hold, a fair election mechanism must rank *A* and *B* in the same relative order in the election outcomes for Γ and Γ'. That is, when the above conditions hold, *A* will be above *B* in the election outcome for profile Γ if and only if *A* is above *B* in the election outcome for profile Γ'. See Figure 4.13.

The IIA requirement is a generalization of the Consistency requirement used in single-winner elections.

Review question: Suppose that *EM* declares the election outcome of *D, A, C, G* for profile Γ shown in Figure 4.1. In order to obey the IIA requirement, must *A* be above *C* in any election outcome for the profile Γ' shown in Figure 4.13?

Does the IIA requirement apply to any other pair of candidates besides (*A*, *C*)?

More on the IIA requirement Requirement IIA is the most technical of Arrow's three requirements, and its implications may be a bit subtle. Because of that, we restate it as:

> **Fact 2:** The IIA requirement means that for any given profile, the relative order of two candidates, say A and B, in an election outcome for the profile, must be determined *solely* by the relative orders of A and B in the n voters preference lists.

Saying this yet another way: Requirement IIA means that as long as the relative order of A and B (in *each* voter's preference list) stays the same over a set of profiles, even if other details of the profiles differ, the election mechanism must report the same relative order of A and B in *all* the outcomes for those profiles – it either reports A above B in all those outcomes or B above A in all those outcomes.

So, when an election mechanism obeys the IIA requirement, an election outcome is really the combination of *separate* election outcomes, one for each *pair* of candidates. With the outcomes from all of those separate "elections", Fact 1 shows how the full election outcome is completely determined.

This view of requirement IIA may sound like it is just a restatement of Fact 1, but don't confuse them. Fact 1 is a mathematical statement about any ordered list, while requirement IIA is a *property* that Arrow *asserted* should be required of any fair election mechanism for ranked-choice voting.

Objections The IIA requirement is the most appropriate and meaningful for ranked-choice elections where a voter's preference list gives an ordering of the candidates but does not provide any measure of the *intensity* of a voter's preferences. For a contrasting situation, see the second Review question on page 79.

Some people object to requirement IIA and have proposed alternatives that they assert are more reasonable requirements for a fair election mechanism, even when restricted to ranked-choice voting. But, as Arrow wrote in 1963 [6]: "... every known electoral system satisfies this condition." Debates about IIA are beyond the scope of this book. For a full defense of requirement IIA, see [6].

Also, you may think that a fair election mechanism should obey some *additional* requirements beyond these three. But, Arrow did not assert that these three requirements (or his original five) alone would be *sufficient* for a fair election mechanism. He only asserted that these requirements are *necessary*, and then proved that it is impossible for any election mechanism to obey even these requirements (when there are at least three candidates).

A formal statement of Arrow's theorem Now that Arrow's three requirements have been explained, we can formally state Arrow's theorem:

Theorem 4.3.1 *There is no election mechanism for ranked-choice voting that can obey the Unanimity, No-dictator, and IIA requirements, provided there are at least two voters and three candidates. Such an election mechanism is impossible.*

Arrow's original published proof Arrow's theorem applies when there are at least two voters and at least three candidates. However, the proof he gave in his original paper in 1950 [4] just addresses the special case of *exactly* two voters and three candidates.[8] Arrow proved his general theorem in a book published the following year, and a somewhat simpler version in the second edition of that book 12 years later [5, 6]. Those proofs are difficult to follow. A succession of simpler proofs of Arrow's general theorem were published over a period of more than 60 years (the most recent proof appearing in 2012 [109]).

In the next section, we will develop a delightful proof of Arrow's general theorem given in a paper by John Geanakopolos [41], published more than 50 years after Arrow's original publications. That proof is vastly simpler than Arrow's proofs of his general theorem in [5, 6].

4.4 A Proof of Arrow's General Theorem

The proof follows a game plan similar to what we saw in the proof of the GS theorem. We assume that an election mechanism, *EM*, obeys the Unanimity and IIA requirements, and then establish that *EM* allows a dictator. So, no election mechanism can obey all three of Arrow's requirements. And similar to the proof of the GS theorem, the proof of Arrow's general theorem examines several scenarios that successively transform one profile into another profile, making observations that show that *EM* allows a dictator.

Note that we do not specify anything about *EM* other than it obeys the Unanimity and IIA requirements. Thus, the examples in the figures only illustrate the ideas – they are not implementing any specific election mechanism. And anything we prove about *EM* is actually a proof about any election mechanism that obeys the Unanimity and IIA requirements.

[8] I have written an exposition on that special case, but not included it in this book. In this book, I directly discuss Arrow's general theorem without first addressing the special case. But, for readers who wish to read my exposition of the special case, it is posted on the Cambridge University Press webpage for this book: https://cambridge.org/9781009349499

Profile Γ_0^A	Voters				
	1	2	3	4 ...	n
	D	C	D	B ...	C
	C	D	C	D ...	D
	B	B	B	C ...	B
	A	A	A	A ...	A

Figure 4.14 Profile Γ_0^A where candidate *A* is at the bottom of every voter's preference list. Any ranking of the other candidates is allowed in Γ_0^A, so the rankings above *A* shown here were selected arbitrarily. Any such profile is said to be of *Type 0*.

4.4.1 Act One: The Start of a Scenario

To start the scenario, we pick one of the candidates to focus on, say candidate *A*. We call it "special". To emphasize that this scenario starts by picking the specific candidate *A* to be special, we call it "Scenario-*A*".

The first profile we create, called Γ_0^A, puts candidate *A* at the *bottom* of every voter's preference list. That is, Scenario-*A* starts with a profile where *A* is every voter's least favorite candidate. All other details of profile Γ_0^A can be *arbitrarily* filled in. So, we know who is at the bottom of every preference list in Γ_0^A (it is *A*), but we don't know who is at the top of any preference list. See Figure 4.14.

For the proof, Γ_0^A is a specific, fixed profile, but since the only constraint on Γ_0^A is that candidate *A* be at the bottom of each voter's preference list, there are many profiles that *could* have been selected. Any such profile is said to be of "Type 0".

Observation F1: For any candidate $X \neq A$, every voter prefers *X* to *A* in Γ_0^A, so by Arrow's Unanimity requirement, *EM* will place candidate *X* above *A* in any election outcome for profile Γ_0^A. It follows that candidate *A* will be at the *bottom* of any such election outcome.

The next part of Scenario-*A* Next, modify profile Γ_0^A by moving candidate *A* from the bottom to the top of the preference list of voter 1 (see Figure 4.15).

After this move, apply *EM* to the modified profile, and if candidate *A* is *no longer* at the bottom of the resulting election outcome, *end* this part of Scenario-*A*. Otherwise, move *A* to the top of the preference list of voter 2; apply *EM* to this modified profile and check whether *A* is still at the bottom of

Profile Γ_0^A	Voters				
	1	2	3	4 ...	n
	A	C	D	B ...	C
	D	D	C	D ...	D
	C	B	B	C ...	B
	B	A	A	A ...	A

Figure 4.15 *Modified* profile Γ_0^A, where candidate *A* is moved to the top of the preference list of voter 1 and all other candidates in that list are moved down one position.

Profile Γ_2^A	Voters				
	1	2	3	4 ...	n
	A	A	A	A ...	C
	D	C	D	B ...	D
	C	D	C	D ...	B
	B	B	B	C ...	A

Figure 4.16 Profile Γ_2^A. The *A*-pivotal-voter is voter 4 in this example. Note that in every voter's list, candidate *A* is either at the top or at the bottom of the list. Also, we have not specified an election mechanism, *EM*, so the declaration of voter 4 as the *A*-pivotal-voter is just to illustrate Scenario-*A*.

the resulting election outcome. *If* candidate *A* is *still* at the bottom of the election outcome, *repeat* this process, with voter 3, 4, ..., (successively moving *A* to the top in the voter's list, applying *EM*, and checking the resulting election outcome), *until* there is an election outcome where *A* is no longer at the bottom of the election outcome. When this occurs, the last voter who moved candidate *A* is called the "*A*-pivotal-voter", and we end this part of Scenario-*A*. The profile at that point is called Γ_2^A. (The "A" in "*A*-pivotal-voter" is to emphasize that voter *A* is special in Scenario-*A*.) See Figure 4.16.

Why must an *A*-pivotal-voter exist? If the moves described for Scenario-*A* continue all the way through the last voter, then candidate *A* will be at the top of *every* voter's preference list, so by the Unanimity requirement, candidate *A* will be at the top of the resulting election outcome, and hence (duh!) *A* will *not* be at the bottom of the outcome. So, there will be a *first* voter (the one we call the *A*-pivotal-voter) in Scenario-*A*, where the move of candidate *A* causes *A* to no longer be at the *bottom* of the resulting election outcome. At that point, the profile is Γ_2^A.

But Wait! – We went from Γ_0^A to Γ_2^A without mentioning a profile Γ_1^A. Let's fix that. We define Profile Γ_1^A as the profile just *before* the *A*-pivotal-voter is

Profile Γ_1^A	Voters				
	1	2	3	4 ...	n
	A	A	A	B ...	C
	D	C	D	D ...	D
	C	D	C	C ...	B
	B	B	B	A ...	A

Figure 4.17 Profile Γ_1^A, just before the move of candidate A resulting in profile Γ_2^A.

discovered in Scenario-*A*, that is just before profile Γ_2^A is created. See Figure 4.17.

Observation F2: By the definition of the *A*-pivotal-voter, the election outcome for profile Γ_1^A will still place candidate *A* at the *bottom* of the election outcome.

But, by the definition of Γ_2^A, *A* is above some candidate, say *Q*, in the election outcome for profile Γ_2^A. In fact, we can establish the stronger fact:

Lemma 4.4.1 *Candidate A must be at the top of the election outcome for profile* Γ_2^A.

Proof For a proof by contradiction, suppose some candidate, say *R*, is above *A*, and hence above *Q* in the election outcome for Γ_2^A.

Now, in every preference list in Γ_2^A, candidate *A* is either at the top or at the bottom of that list (see again Figure 4.16), so in every preference list, either *both R* and *Q* are below *A* or both *R* and *Q* are above *A*. Consider next a new profile, Γ', created from Γ_2^A by exchanging the positions of *Q* and *R* in *every* preference list where *R* is *above Q*. The result is that *Q* is above *R* in every voter's list in Γ', and so by the Unanimity requirement, *Q* must be above *R* in the election outcome for Γ'. See Figure 4.18.

Since, in any voter's list, *R* and *Q* are either both above *A* or both below *A*, the relative order of *A* and *R* will be the same in Γ' and Γ_2^A. We assumed that *R* is at the top of the election outcome for Γ_2^A (and hence above *A* in that outcome), so *Fact 2* dictates that *R* must be above *A* in the election outcome for Γ'. But, the relative order of *A* and *Q* is also the same in both Γ' and Γ_2^A, and *A* is above *Q* in the election outcome for Γ_2^A, so *Fact 2* dictates that *A* must be above *Q* in the outcome for Γ'. Then, by transitivity, it follows then that *R* must be above *Q* in the election outcome for Γ'. But that is a contradiction, because we previously established that *Q* must be above *R* in the election outcome for Γ'.

Γ_2^A							Γ'						
1	2	3	4	.	.	n	1	2	3	4	.	.	n
A	A	A	A	.	.	R	A	A	A	A	.	.	**Q**
Q	.	.	.	R	.	.	Q	.	.	.	**Q**	.	.
.	R	.	Q	.	Q	.	.	**Q**	.	Q	.	Q	.
R	Q	Q	.	.	.	Q	R	**R**	Q	.	.	.	**R**
.	.	R	.	Q	R	.	.	.	R	.	**R**	R	.
.	.	.	R	A	A	A	.	.	.	R	A	A	A

Figure 4.18 Profile Γ' is created from Γ_2^A by reversing the positions of R and Q in any list where R is above Q. The entries that changed are shown in bold in Γ'. The result is that Q is above R in every list in Γ', and hence by the Unanimity requirement, Q must be above R in the election outcome for Γ'. Note that in every list in Γ_2^A, candidate A is either above both R and Q or below both. Hence, after reversing the positions of R and Q in some lists, the relative orders of A and R, and of A and Q, in any voter's list, are the same in both Γ_2^A and Γ'.

Recapping: Starting from the supposition that candidate A is *not* above all candidates in the election outcome for Γ_2^A, we have come to the contradiction: R is above Q, and Q is above R, in the election outcome for Γ'. Therefore, the supposition must *not* be true, and so there can be *no* such candidate R that is above A in the election outcome for Γ_2^A – candidate A must be above all other candidates in that election outcome. ■

4.4.2 Act Two: Onward with Scenario-A

We will use Lemma 4.4.1 in a bit, but first we examine some modifications of profile Γ_2^A. *Arbitrarily* select two of the m candidates other than candidate A. Since the choices are arbitrary (as long as A is not selected), we will just call them X and Y. But note that once selected, the identities of X and Y must remain *fixed* in the argument until after the statement of Theorem 4.4.3.

Modify profile Γ_2^A by replacing the preference list for the *A-pivotal-voter* with a new preference list where the selected candidate X is *above* candidate A and the selected candidate Y is *below* A. There is no restriction on where candidates A, X, and Y can go in the preference list of the A-pivotal-voter as long as the relative order of the candidates is X above A and A above Y. Once those three candidates are placed, the remaining candidates can be put arbitrarily into the available positions in the list of the A-pivotal-voter. This profile is called Γ_3^A. See Figure 4.19.

Observation $F3$: In both profiles Γ_1^A and Γ_3^A, candidate A is above candidate X in the preference list of every voter to the *left* of the A-pivotal-voter, and

Profile Γ_3^A	Voters				
	1	2	3	4 ...	n
	A	A	A	X ...	C
	D	C	D	B ...	D
	C	D	C	A ...	B
	B	B	B	Y ...	A

Figure 4.19 Profile Γ_3^A. Recall that the A-pivotal voter is voter 4 in this example. Candidates X and Y are C and D in this illustration, but we don't know which one is C and which is D.

below X in the preference list of the A-pivotal-voter and every voter to its *right*. This follows by the process that transformed Γ_1^A to Γ_3^A. Hence, in the preference list of any voter, candidates A and X are in the same relative order in profiles Γ_1^A and Γ_3^A.

Observation $F3$ is useful because we saw that candidate A is at the bottom of the election outcome for profile Γ_1^A, so X is above A in that election outcome. Hence, by Fact 2, candidate X must also be above A in the election outcome for profile Γ_3^A. Similarly, in the preference list of any voter, candidates A and Y are in the same relative order in profiles Γ_2^A and Γ_3^A. But, by Lemma 4.4.1, A is at the top of the election outcome for Γ_2^A, so by Fact 2, candidate A must be above Y in the election outcome for profile Γ_3^A.

Recapping: We have deduced that candidate X is above candidate A, and that A is above candidate Y in the election outcome for profile Γ_3^A. Hence, by transitivity, candidate X is above candidate Y in the election outcome for profile Γ_3^A. We will use this fact shortly.

4.4.3 Back to Scenario-*A*

As a warm-up to the final modifications in Scenario-A, arbitrarily place candidate A *anywhere* in the A-pivotal-voter's preference list. We call the resulting profile Γ_4^A. Note that Γ_4^A could be identical to Γ_3^A.

To be more specific, if candidate A is moved down from a position k to a lower position k' in the list, then the candidates who were at positions $k+1$ through k' in Γ_3^A are moved one position higher. Similarly, if A is moved up from position k to a higher position k', then the candidates who were at positions k' through $k-1$ are each moved *down* by one position. See Figure 4.20.

Observation $F4$: Candidate X must be above candidate Y in the election outcome for profile Γ_4^A.

Profile Γ_4^A	Voters				
	1	2	3	4 ...	n
	A	A	A	X ...	C
	D	C	D	**A ...**	D
	C	D	C	**B ...**	B
	B	B	B	Y ...	A

Figure 4.20 Profile Γ_4^A created from Γ_3^A (shown in Figure 4.19) after moving candidate A in the list of voter 4, the A-pivotal-voter in the running example.

To see this, note that in profile Γ_4^A, the relative order of X and Y in the preference list of any voter j is the same as the relative order of X and Y in the preference list of voter j in profile Γ_3^A. But, we just established that candidate X is above Y in the election outcome for profile Γ_3^A, so by *Fact 2*, this must also be true for profile Γ_4^A.

4.4.4 The Last Modifications

For our final modifications, (possibly) change profile Γ_4^A by placing candidate A *anywhere* in *any* voter's preference list. Call the resulting profile Γ_5^A. In any voter's preference list, the relative order of candidates X and Y in profile Γ_5^A is the same as in profile Γ_4^A, so by exactly the same reasoning used to establish *Observation* F4, candidate X must be above candidate Y in the election outcome for profile Γ_5^A.

Regrouping So far, we have established:

Theorem 4.4.1 *Candidate X is above candidate Y in the* preference list *of the A-pivotal-voter in profile* Γ_5^A *and also in the* election outcome *for profile* Γ_5^A.

Weak, but growing, evidence Recall that the choice of candidates X and Y was arbitrary as long as neither was candidate A. Furthermore, the relative order of X and Y in every list other than the list of the A-pivotal-voter is completely arbitrary – their order was arbitrary in Γ_1^A, and no modifications leading to Γ_5^A changed their relative order, except (possibly) in the list of the A-pivotal-voter. So in Γ_5^A, the relative order of X and Y is arbitrary in every preference list other than in the list of the A-pivotal-voter. And yet, we know for sure that X is above Y in the *election outcome* for profile Γ_5^A. So, the relative order of X and Y in the list of the A-pivotal-voter is what determines that X will be above Y in that election outcome. This gives some (still weak) evidence that the A-pivotal-voter may be a dictator. But we need to establish more.

Toward that end, consider what would happen if we go back to the start of the Scenario-*A*, but instead of starting with profile Γ_0^A, we start with some *other* Type 0 profile – call it $\overline{\Gamma_0^A}$.

If we start the new scenario with profile $\overline{\Gamma_0^A}$ and follow the same process as in Scenario-*A*, successively moving candidate *A* to the top of the preference lists of voter 1, voter 2, ..., there will be a *first* voter, say *voter i*, with the property that after their move, candidate *A* will no longer be at the bottom of the election outcome. From there, if we continue to follow the same process as in Scenario-*A*, and select the same candidates for *X* and *Y*, we will obtain a profile, call it $\overline{\Gamma_5^A}$, where candidate *X* is *above* candidate *Y* in the *preference list* of *voter i*. And (following the logic that established Theorem 4.4.1) *X* will be above *Y* in the election outcome for $\overline{\Gamma_5^A}$.

Is this progress or a dead-end? We have now established some evidence that *voter i* may be a dictator. But, the evidence that voter *i* is a dictator is exactly the same as our prior evidence that the *A*-pivotal-voter is a dictator. By the very definition of a dictator, there can never be *two* of them, so, we must conclude that *either* there is *no* dictator *or* that the *A*-pivotal-voter and voter *i* are the *same* voter. Which one is right?

Observation *F*5: The *A*-pivotal-voter (defined in Scenario-*A*) and voter *i* (defined in the new scenario), *must* be the *same* voter.

To see this, remember that profile Γ_2^A is the first profile in Scenario-*A* where *A* is above some other candidate, say candidate *Q*, in the election outcome for that profile. Now suppose voter *i* is to the *right* of the *A*-pivotal-voter. Then in the new scenario, after voters 1 through the *A*-pivotal-voter move candidate *A*, *A* will be above candidate *Q* in exactly the same preference lists as in profile Γ_2^A. So, by Fact 2, at that point in the new scenario, *A* would be above *Q* in the election outcome, and so the new scenario should have stopped before reaching voter *i*. Hence, voter *i* cannot be to the right of the *A*-pivotal-voter. The same kind of argument applies if voter *i* is to the *left* of the *A*-pivotal-voter (i.e., where the *A*-pivotal-voter is to the right of voter *i*).

> **Review question:** Write out in full the argument that the *A*-pivotal-voter cannot be to the right of voter *i*.

Hence, the *A*-pivotal-voter and voter *i* are the *same* voter, and we have established the following extension of Theorem 4.4.1:

Theorem 4.4.2 *For* any *profile* $\overline{\Gamma_5^A}$ *that can be obtained from* some *Type 0 profile by following the process described in Scenario-A, candidate X must be* *above candidate Y in the* preference list *of the A-pivotal-voter, and candidate* X must *also be above candidate Y in the election outcome for profile $\overline{\Gamma_5^A}$. The relative order of X and Y in any list other than the list of the A-pivotal-voter is irrelevant.*

Theorem 4.4.2 would be strong evidence that the A-pivotal-voter is a dictator, depending on how many (and which) profiles can play the role of $\overline{\Gamma_5^A}$. So we ask: Under what conditions can a profile Γ be $\overline{\Gamma_5^A}$?

We answer the question Consider what parts of any $\overline{\Gamma_5^A}$ are constrained and what parts are arbitrary. The only constraint on a Type 0 profile is that candidate A is at the bottom of each preference list. All other features of a Type 0 profile are arbitrary. Then, following the details of Scenario-A from a Type 0 profile to the resulting profile $\overline{\Gamma_5^A}$, we see that the only preference list that is in any way constrained in $\overline{\Gamma_5^A}$ is the list of the A-pivotal-voter. The preference list of every other voter is completely *arbitrary*, that is, it could be *any* arrangement (permutation) of the m candidates.

Further, the only constraint on the preference list of the A-pivotal-voter in profile $\overline{\Gamma_5^A}$ is that candidate X is above candidate Y in that list. Then, we have

Observation $F6$: For any profile Γ where candidate X is above Y in the preference list of the A-pivotal-voter (and all other details are arbitrary), there is a Type 0 profile, $\overline{\Gamma_0^A}$, and a scenario that implements the same process as in Scenario-A, where profile $\overline{\Gamma_0^A}$ is the first profile, and profile Γ is last profile, in the scenario.

Hence, anything we proved earlier about the specific profile Γ_5^A in Scenario-A holds for any profile where candidate X is above candidate Y in the preference list for the A-pivotal-voter. In particular, we have

Theorem 4.4.3 *For* any *profile Γ where candidate X is above candidate Y in the* preference list *of the* A-pivotal-voter, *candidate X must be* above *candidate Y in the election outcome for* Γ.

Now, we remember that the choices of candidates X and Y were arbitrary, as long neither were candidate A, so Theorem 4.4.3 can be extended to:

Theorem 4.4.4 *For* any *selection of two different candidates, where neither candidate is candidate A, and for any profile, Γ, if the first selected candidate is above the second selected candidate in the* preference list *of the* A-pivotal-voter, *then the first selected candidate must be* above *the second selected candidate in the election outcome for profile* Γ.

Profile Γ_0^B	Voters				
	1	2	3	4 ...	n
	D	C	D	A ...	C
	C	D	C	D ...	D
	A	A	A	C ...	A
	B	B	B	B ...	B

Figure 4.21 The Type 0 profile, Γ_0^B, obtained from Profile Γ_0^A shown in Figure 4.14. Profile Γ_0^B is created by exchanging the positions of candidates A and B in Profile Γ_0^A.

Almost there Theorem 4.4.4 is very strong evidence that the A-pivotal-voter is a dictator. In fact, it would satisfy the definition of dictator, except that Theorem 4.4.4 does not refer to any pairs of candidates that contain candidate A.

Review question Suppose that Theorem 4.4.4 did *not* include the phrase "where neither candidate is candidate A". Explain why that modified theorem would imply that the A-pivotal-voter is a dictator. Use the definition of a dictator and Fact 1.

4.4.5 The Final Act: Another Renaming Argument

How can we include candidate A in Theorem 4.4.4? Well, what really is in a name? Couldn't we run a new scenario where we don't start the scenario with candidate A? Hint: Yes we can! And what that means is that we would again prove a theorem like Theorem 4.4.4, but where A could be one of the two candidates referred to as X or Y in Scenario-A. Hopefully, after having seen a similar renaming argument in the discussion of the GS theorem, you are comfortable with this kind of argument, but to be sure, let's run through it explicitly.

Go back to the start of Scenario-A, and rerun the scenario, but this time instead of choosing candidate A to be special, choose another candidate, call them B. The resulting scenario will be called Scenario-B. In particular, take profile Γ_0^A, defined for Scenario-A, and exchange the positions of candidates A and B in each voter's preference list. We call this modified profile Γ_0^B. See Figure 4.21.

So, in every voter's preference list in profile Γ_0^B, candidate B is at the bottom, and candidate A is where candidate B was in profile Γ_0^A. Then, profile Γ_2^B is created by successively moving candidate B from the bottom to the top of the preference lists of voter 1, voter 2, ..., as in Scenario-A, until there is an election outcome where B is no longer at the bottom of the election outcome list. When this occurs, the voter who just moved B is called the "B-pivotal-voter", and we end this part of Scenario-B.

Now we ask the key question: What is the relationship of the A-pivotal-voter to the B-pivotal-voter? By the details of Scenario-A and Scenario-B, each profile in the series Γ_0^A through Γ_2^A is identical to the corresponding profile in the series Γ_0^B through Γ_2^B, except that the positions of candidates A and B have been exchanged. So, profiles Γ_2^A and Γ_2^B differ only by the exchange of candidates A and B in each voter's preference list. Otherwise, profiles Γ_2^A and Γ_2^B are identical. Hence, the first point in Scenario-A where A is no longer at the bottom of the election outcome is also the first point in Scenario-B where B is no longer at the bottom of the election outcome. That is, the A-pivotal-voter and the B-pivotal-voter are the same voter.

From profile Γ_2^B, we can complete Scenario-B by making each profile in Scenario-B identical to the corresponding profile in Scenario-A, except that again the positions of candidates A and B are exchanged. So everything we previously proved about the A-pivotal voter now applies to the B-pivotal voter.

The result is that we can replace A with B everywhere in Theorem 4.4.4 and get a new theorem that holds for all pairs of candidates X and Y, as long as neither X nor Y is candidate B. But, X or Y could be candidate A. Then, together with Theorem 4.4.4, we have covered *almost* all pairs of candidates. The only pair that is not covered by these two theorems is the pair $\{A, B\}$. That is, the assignments $X = A$, $Y = B$ and $X = B$, $Y = A$ are not covered. What to do?

The last scenario Well, we assumed that there are at least three candidates, so there is one, called C, which is neither A nor B. Note that the ordered pairs (B, C) and (C, B) were covered in Theorem 4.4.4, and that the ordered pairs (A, C) and (C, A) were covered in its modification above.

So rerun Scenario-A again, but this time select candidate C to be special. The resulting scenario is called Scenario-C. We will now find a C-pivotal-voter, and by the above reasoning, it must be the same voter as the A-pivotal-voter (and the B-pivotal-voter). Modifying Theorem 4.4.4 by replacing A with C everywhere, we get another theorem where the assignments $X = A$, $Y = C$ and $X = C$, $Y = A$ are covered. Now, we have covered all pairs and this finally gives us:

Theorem 4.4.5 *(The* **Dictator** *Theorem) For* any *selection of two different candidates, and for any profile,* Γ, if *the first selected candidate is above the* *second selected candidate in the preference list of the* A-pivotal-voter *in* Γ, then *the first selected candidate must also be above the second selected candidate in the election outcome for profile* Γ.

Theorem 4.4.5, along with Fact 1, establishes that the A-pivotal-voter really *is* a dictator, finishing the proof of Arrow's General Impossibility Theorem. ■

Such a profound theorem, and such a delightful proof! And, only careful, logical reasoning (but, a fair amount of it) was ever used – no deep math needed.

4.5 Exercises

1. Explain why the method for determining the order of an unknown list, described in footnote 7, is correct.
2. Fact 1 says that the relative orders of all the pairs of candidates in an ordered list uniquely determine the order of the list. Now, consider an example of three candidates $\{A, B, C\}$ where the relative orders of all the pairs of candidates are: A above B, B above C, and C above A. These three relative orders do not determine an ordering of the candidates $\{A, B, C\}$. Why not? Does this contradict Fact 1? Explain.
3. Show that Arrow's theorem holds when there is only a *single voter* and any number of candidates. This is really just an exercise in using the definitions, but the definitions do actually lead to Arrow's theorem in the case of only a single voter.
4. Consider again the election mechanism described in the Review question on page 79 and the profile Γ in Figure 4.1.

   Show that this proposed election mechanism does *not* satisfy requirement IIA. This can be done by giving a profile where in each voter's preference list, the relative order of A and C is the same as in Γ, but with the proposed election mechanism, C would be ranked above A in the election outcome.

   We said earlier that requirement IIA was most appropriate for ranked-choice voting where rankings do not provide information about the *intensity* of a voter's preferences. In what way does the election mechanism on page 79 implicitly infer intensity information?
5. Arrow's original paper [4] only proves his impossibility theorem for the case of *exactly* two voters and three candidates. It did not give a proof of Arrow's general theorem.

One might argue that proving Arrow's theorem for the case of only two voters and three candidates is the "moral equivalent" of Arrow's general theorem (with more than two voters and at least three candidates), since it already shows that his three requirements are not always attainable in any election mechanism.

On the other hand, if we only have Arrow's original proof, it would still leave the possibility that there *is* a fair election mechanism when there are *more* than two voters and three candidates. In that case, Arrow's theorem would still be correct, but its implications would be less compelling.

What do you think?

6. The proof presented here for Arrow's General Theorem does not mention the possibility of an election mechanism declaring a tie. Show how to incorporate such a possibility into the proof. Does the statement of Arrow's theorem need to be changed, or just the proof?
7. The exposition given here for Arrow's General Theorem follows the first of three proofs developed in [41], giving more complete and explicit explanations for each step – the exposition in [41] takes a little over one page, including one figure! A later paper [109] claims to have found an even briefer, more direct *proof* of Arrow's General Theorem. The exposition is a bit shorter than the one in [41].

   While the proof in [109] and the first proof in [41] differ in some details, I don't find those differences of fundamental importance, and I found it much harder to follow. Perhaps it is just a briefer exposition, rather than a new proof.

   If you are really motivated, read the proof in [109] and then write an expanded exposition of it, with much more complete explanations of each step, similar to the way I wrote the proof from [41]. Then, compare your expanded exposition with the one in this chapter to see if you find significant differences in the proofs. Let me know what you find out.

5
Clustering and Impossibility

5.1 The Importance of Clustering

Clustering in the *analysis of data* is one of the most critical operations in many application areas, particularly in the growing fields of *machine learning* and *data science*. There are hundreds of methodological papers on proposed ways to do clustering and tens of thousands of papers where clustering methods have been applied to different datasets to try to extract important insights from the data.

5.1.1 Clusters and Clustering

In a large set of data $\mathcal{S}$, a *cluster* is a *subset* of $\mathcal{S}$ where the elements in the subset are highly related to each other and much less related to the data outside of the subset. Then, *clustering* is the task of dividing the elements of $\mathcal{S}$ into several nonoverlapping clusters. We want a clustering that gives meaningful insight into the substructure and possible communities in the data. Figure 5.1 (panels a and b) illustrates these definitions with points on the plane where, in this example, the relatedness of two points is measured by their straight-line distance from each other. Straight-line distance is usually called *Euclidian distance*.

More generally, for each pair of elements (i, j) in set $\mathcal{S}$, there is a distance, denoted $D(i, j)$, between elements i and j. We call D a *distance function*. Note that in general, distances need not have any relation to a Euclidian distance.

Two real clustering studies To further illustrate the importance and challenges of meaningful clustering, we discuss two actual clustering studies.[1]

[1] The discussion of these two examples is copied (with small modifications) from [48], with permission from the author.

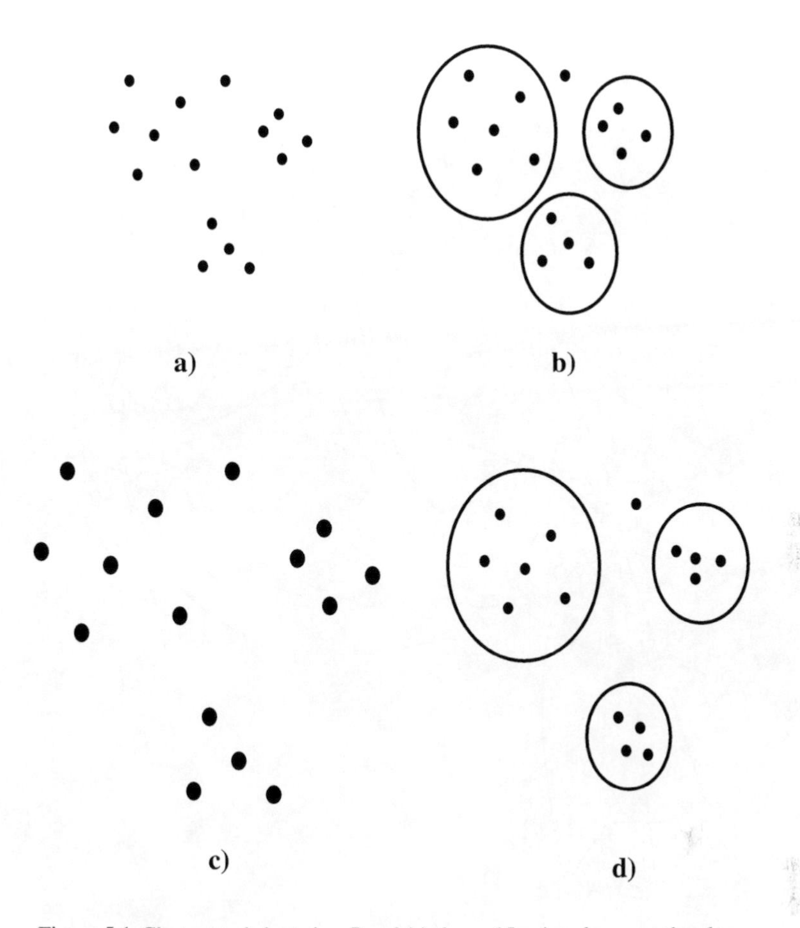

Figure 5.1 Clusters and clustering. Panel (a) shows 15 points drawn on the plane. The set of distances D between any two points is just the straight-line (Euclidean) distance, making it sensible to cluster the points by eye. Panel (b) shows a plausible clustering into four clusters, including one cluster just consisting of a single point. An alternative clustering might add that point to one of the two neighboring clusters. Panel (c) shows the same points where all the distances have been proportionally scaled up, so the relative distances remain the same as before. Panel (d) shows altered distances (not simple proportional scaling) that are *consistent* with the distances and clustering shown in panel (b).

First example Consider the network shown in Figure 5.2. This is the famous example of Zachary's Karate Club [110], which is widely discussed in the clustering and community-finding literature [44]. Each node in this network represents one of the 34 members of the club. Each edge (line) in the network connects two nodes in the network and represents the fact that the two

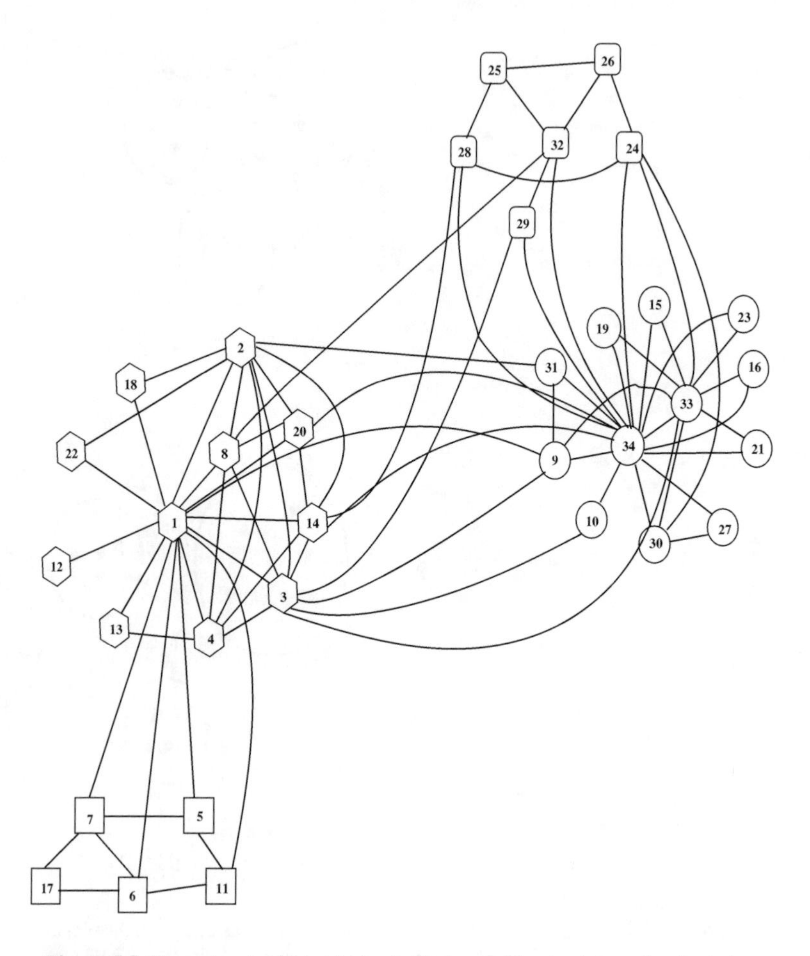

Figure 5.2 The network of friendships in Zachary's Karate club, and a clustering of the 34 members into four clusters found by the Louvain algorithm [48, 78]. Each edge (line) connecting two nodes in this graph indicates that the two people (represented by two nodes) are good friends. If there is no edge between two specific nodes, then the two people represented by those nodes are not good friends. The length of an edge has no meaning. The nodes in each cluster are identified by the geometric shape used for the node. The shapes are circles, squares, hexagons, and squares with rounded corners. This figure and Figure 5.3 are plagiarized from [48] without written permission from the author or publisher. Don't tell them.

members represented by those two nodes are good friends. In this example, the "relatedness" of two members is a *binary* relation, rather than a numerical relation. The relatedness here simply indicates whether or not two members are good friends.

The story goes that the karate club was being disbanded and needed to be divided into several subsets of members, where in each subset the members were (generally) good friends of each other, and generally less friendly with members outside of that subset. That was a *clustering* problem, since the goal of clustering is to identify subsets of elements in a large dataset that are highly *related* to each other and much less related to elements outside of the cluster. Each such subset is a *cluster*. Figure 5.2 shows a clustering of the nodes in Zachary's Karate club into four clusters. This is the clustering that is obtained from one of the most widely used community detection algorithms. Community detection is a particular application of clustering.

Review question At first glance, the four clusters depicted in Figure 5.2 seem sensible, but a more quantitative analysis makes that less clear. Count how many edges there are inside clusters, and how many edges there are crossing between clusters, in Figure 5.2.

Now suppose we move nodes 24 and 28 from the cluster they are in to the cluster whose nodes are represented by circles. How do the edge counts change after those moves?

How many edges are there between nodes 24 and 28 and other nodes in the same cluster they are in, both before and after the moves of nodes 24 and 28?

Given these counts, do you think that the clustering shown in Figure 5.2 is actually the most meaningful or natural one?

Draw out the clustering where nodes 24 and 28 are moved to the cluster represented by circles. Stare at that clustering for a while. Does it seem a more natural or informative clustering than the one shown in Figure 5.2?

Second example A recent article [36] involving a network called a food web received considerable attention in the popular press, including *The New York Times*. The feeding and co-feeding behaviors of 13 mountain lions (with GPS tracking collars) in Wyoming were observed using several hundred motion-triggered cameras placed at sites where mountain lions had previously killed and eaten large animals. Mountain lions are thought to be solitary animals that do not generally share food or cooperate with other mountain lions (except to mate). The cameras caught 48 occurrences of individual lions with a kill that was larger than the lion could eat in one meal, and the subsequent arrival of another lion. The data of interest was whether the first lion shared the carcass with the second lion or fought with it.

The recorded data were organized as a network, where each node represented one of the 13 lions, with a *directed* edge from the node for the first

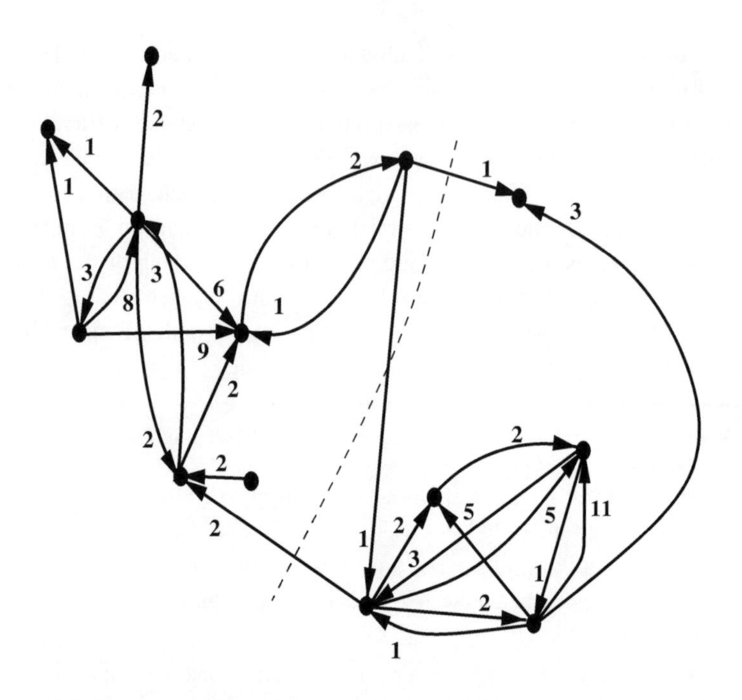

Figure 5.3 The lion food-sharing network (adapted from Figure 1 in [36]). Each node represents one of the 13 lions studied. An edge from a node i to a node j indicates that lion i killed prey and shared the kill with lion j. The number on the edge indicates the number of such interactions. Any missing directed edge from node i to node j means that lion i did not share any kill with lion j. The dashed line separates the two communities (clusters) identified in [36].

lion to the node for second lion, *if* the carcass *was* shared. A representation of that network is shown in Figure 5.3. To extract biological meaning from the network, the researchers analyzed the network

> to quantify the extent to which the network was composed of distinct communities (also called "clusters") within which many edges occurred, and between which few edges occurred. [36]

The result, as can be seen in Figure 5.3, is a division into two distinct food-sharing communities, separated by the dashed line. The community (cluster) membership of most of the lions is clear, but the membership of the upper-right lion might be debated. Much is unknown about the communities, for example, how they were established, but knowing that such clusters exist, and which lions are members, is an important and necessary start. More details on the lions (e.g., their sex, weight, and age) are given in [36].

Review question Referring to the clustering shown in Figure 5.3, why might the community membership of the upper-right lion be debated?

The key questions Having seen an abstract example of clustering, and two real examples of clustering, the key questions are:

What makes clusterings good or informative?
How do we find them?

Despite many proposed clustering algorithms, and a huge number of empirical studies where clusters in datasets have been suggested, there has been very little research on fundamental and foundational answers to these key questions. One of the few such efforts is discussed next.

In this short chapter In this chapter, we discuss an *axiomatic* approach to defining what a good cluster *algorithm* (or mechanism) should be and then prove a theorem that shows that it is *impossible* to devise such an algorithm. This axiomatic approach and the resulting impossibility results follow the intellectual tradition initiated by Arrow and his impossibility theorem, discussed Chapter 4.

The specific proof we present will also show that with somewhat more *realistic* axioms, it is still impossible to develop a clustering algorithm that satisfies all of the axioms. However, that result is not as informative as it seems at first, since it is made invalid by a small change in the model. Thus, this chapter illustrates the interplay of *models* and *theorems*, an issue that also arose in our discussion of Bell's and Arrow's impossibility theorems.

5.2 Clustering Axioms and Impossibility

We use F to denote a clustering algorithm and define $F(\mathcal{S}, D)$ as the clustering created by algorithm F when given the set of elements $\mathcal{S}$ and the set of distances D between the pairs of elements in $\mathcal{S}$. We can refer to D either as a set of distances between pairs of elements in $\mathcal{S}$ or as a *distance function* of pairs of elements in $\mathcal{S}$. Recall that the distance function D is very general and need not have any relation to pairwise *Euclidean* distances of the elements.

What is a good clustering? Now, the distances (given by D) can vary, and the resulting clustering $F(\mathcal{S}, D)$ may change as the distances change. So, for a fixed set of elements, $\mathcal{S}$, and a fixed clustering algorithm, F, the resulting clustering $F(\mathcal{S}, D)$ is a *function* of D. With these definitions, we can ask:

> What properties should a good clustering algorithm, F, have? Equivalently, what properties should a good clustering function, $F(\mathcal{S}, D)$, have?

5.2.1 Kleinberg's Axioms

Jon Kleinberg, in 2002 [57], proposed three *axioms* (or *requirements*) that a good clustering algorithm, F, should obey,[2] and then proved that *no* clustering algorithm can simultaneously obey all of them – it is *impossible*.

The axioms and their meanings Kleinberg's three axioms are as follows:

1. **Scale invariance:**
   If two sets of distances differ only by a *multiplicative constant*, then the resulting clusterings given by a good clustering algorithm F should be the same.

   In more detail, suppose D and D' are two sets of distances between pairs of elements of set $\mathcal{S}$. If there is a constant number α where $D'(i,j) = \alpha \times D(i,j)$ for each pair of elements (i,j) in $\mathcal{S}$, then clustering $F(\mathcal{S}, D)$ should be the same as clustering $F(\mathcal{S}, D')$.

   The intuition for this axiom is that good clustering algorithms should *not* be influenced by the *absolute* values of the distances given to them but only by the *relative* values of the distances. Scaling of the distances should have no affect on the resulting clustering. In Figure 5.1 (panel c), the distances are proportionally scaled up from the distances in panel (a), and it is reasonable that good clusterings should be the same for both sets of distances.

   The scale-invariance axiom is generally accepted as reasonable.
2. **Richness:**

   A good clustering algorithm, F, should have the property that for *any* clustering $C(\mathcal{S})$ of set $\mathcal{S}$, there is some set of distances, D, such that $F(\mathcal{S}, D) = C(\mathcal{S})$. In other words, for any way of dividing up the elements of $\mathcal{S}$ into nonoverlapping clusters, $C(\mathcal{S})$, there will be some set of distances

[2] As we will see, there is disagreement about the appropriateness and robustness of Kleinberg's three axioms, but the paper was widely read because despite the importance of clustering, there were very few theoretical results concerning the foundations and fundamental properties of clustering. Kleinberg's axioms and paper opened the door to more research on foundational issues of clustering.

D (maybe not realistic ones in a given application) such that algorithm F, given $\mathcal{S}$ and D, will create exactly those clusters.

Note that because of the Richness Axiom, there must be an extreme set of distances, D, where algorithm F puts every element in $\mathcal{S}$ into its own, *separate* cluster. We call that clustering the *anti-social* clustering. Symmetrically, there must be an extreme set of distances, D', where algorithm F puts all of the elements into one *single* cluster. These two extreme clusterings are in addition to all of the other clusterings created by algorithm F for less extreme distances.

The richness axiom has been criticized for allowing extreme clusterings, something that is not allowed by many clustering algorithms. However, the richness axiom does not require that those extreme clusterings would be created by "reasonable" distances, that is, ones that would ever be encountered in practice.

3. **Consistency:**
   Consider a clustering $F(\mathcal{S}, D)$ for a set of distances, D, and let D' be any different set of distances, $D' \neq D$. Then D' is called *consistent* with the pair $(D, F(\mathcal{S}, D))$, if $D'(i, j) \geq D(i, j)$ for every pair of elements (i, j) that are in *different* clusters of $F(\mathcal{S}, D)$ and $D(i, j) \leq (D(i, j)$ for every pair of elements (i, j) that are in the *same* cluster of $F(\mathcal{S}, D)$.

   Stated differently, D' is consistent with the pair $(D, F(\mathcal{S}, D))$ if, when changing distances from D to D', the distances *between* clusters of $F(\mathcal{S}, D)$ only *increase* or stay the same, and the distances *inside* clusters of $F(\mathcal{S}, D)$ only *decrease* or stay the same. Figure 5.1 (panel d) gives an example of a set of distances, D', that is consistent with the pair $(D, F(\mathcal{S}, D))$ from panel b).

   A clustering algorithm, F, is said to be *consistent* if for any two set of distances, D and D', when D' is consistent with the pair $(D, F(\mathcal{S}, D))$, then the clustering $F(\mathcal{S}, D')$ is the *same* as the clustering $F(\mathcal{S}, D)$.

   That is, starting from a clustering $F(\mathcal{S}, D)$, if the distances between pairs of elements inside the same cluster stay the same or *decrease*, and the distances between pairs of elements in different clusters stay the same or *increase*, then the clustering does *not* change.

   Note that these uses of the word "consistency" are different from its use in our discussion of the GS theorem in Chapter 4.

Consistency is a very strong and highly disputed axiom. For example, in Figure 5.1 (panel b), if the upper right cluster, along with the cluster containing only a single point, were pulled a million miles to the right of the upper left cluster, it would then be sensible to merge the single point cluster into the upper right cluster. This is already visually suggested in Figure 5.1 (panel d). But the consistency axiom does not allow such mergings.

Later in this chapter, we discuss alternatives to the consistency axiom that are less extreme and still lead to impossibility. But first, we present the main impossibility result from [57].

Theorem 5.2.1 *There is no clustering algorithm that simultaneously obeys the axioms of scale invariance, richness, and consistency.*

Theorem 5.2.1 was first proved in [57], but our discussion follows the simpler proof presented in the dissertation of Margarita Ackerman [2].

Proof of Theorem 5.2.1: Consider two clusterings: the antisocial clustering, we call C_0, which puts each element in its own cluster; and C_1, which is any other clustering different than C_0. By the *richness* axiom, each clustering C is associated with some set of distances, D, where C is the clustering $F(\mathcal{S}, D)$ created by algorithm F given the distances D. So, there must be some set of distances, D_0, which leads F to create clustering C_0; and also some set of distances, D_1, which leads F to create clustering C_1.

We define $\max(D_0)$ be the *largest* number in the distances D_0. Next, let α be some number where $\alpha \times D_1(i,j)$ is greater than $max(D_0)$, for each pair of elements (i,j) in $\mathcal{S}$. Then, define a new set of distances

$$D'(i,j) = \alpha \times D_1(i,j),$$

for each pair of elements i,j in $\mathcal{S}$. Then, $D'(i,j) > \max(D_0) \geq D_0(i,j)$, for each pair i,j. For example, see Figure 5.4.

Algorithm F will create a clustering, $F(\mathcal{S}, D')$, for distances D'. What will that clustering be? We first argue that it must be C_1, and we next argue that it must be C_0.

Certainly, it must be C_1 Since clustering C_1 is generated by F for distances D_1, and the distances in D' are all created by multiplying the distances in D_1 by a single number α, the *scale-invariance* axiom requires that algorithm F create the same clustering, C_1, for the distances in D'. That is, $F(\mathcal{S}, D') = F(\mathcal{S}, D_1)$.

But No! it must be C_0 In C_0 every element in $\mathcal{S}$ is in its own, separate cluster, so that each (i,j) is a *between-clusters* pair. Then, since $D'(i,j) > D_0(i,j)$ for each pair (i,j), distance D' is consistent with the pair (D_0, C_0). So, by the *consistency* axiom, algorithm F must create the same clustering C_0 for the distances in D'. That is, $F(\mathcal{S}, D') = F(\mathcal{S}, D_0)$.

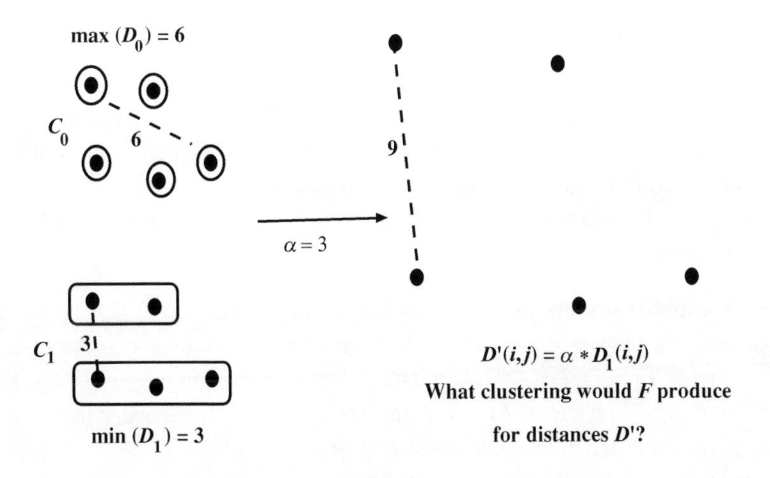

Figure 5.4 Clustering $C_0 = F(\mathcal{S}, D_0)$ is the antisocial clustering, and clustering $C_1 = F(\mathcal{S}, D_1)$ is some other clustering. The largest D_0 distance is assumed to be 6, and the smallest D_1 distance is assumed to be 3. If α is chosen to be 3, then $D'(i, j) = 3 \times D_1(i, j) > \max(D_0) \geq D_0(i, j)$, for each pair of elements (i, j) in $\mathcal{S}$. The points on the right side of the figure illustrate the distances D' (although not to scale). What clustering will algorithm F produce for distances D'?

Thus, one line of reasoning leads to the conclusion that the clustering that F will produce for D' will be C_1, while the other line of reasoning leads to a different conclusion, namely that the clustering for D' will be C_0. Since clusterings C_0 and C_1 are different, this is a contradiction. In reaching this contradiction, we only assumed Kleinberg's three clustering axioms, and hence we conclude that it is *impossible* to have an algorithm that creates clusterings from distances and obeys all three requirements: *scale invariance*, *richness*, and *consistency*. ■

Review question Would the proof of Theorem 5.2.1 work if we make C_1 the clustering where all elements of $\mathcal{S}$ are in a single cluster?

Review question Suppose we change the richness axiom to: There must be some set of distances such that algorithm F produces the antisocial clustering for those distances; and there must be some set of distances such that the algorithm produces at least one clustering other than the antisocial clustering.

Can there be an algorithm that obeys this modified Richness axiom, along with the scale invariance and consistency axioms?

5.2.2 More Realistic Axioms

The consistency axiom is pretty restrictive, and it is not clear why we should expect any clustering algorithm to obey it. But, there are two more realistic consistency-related axioms that have been studied, where both still lead to impossibility results. One axiom, called *refinement consistency*, was introduced in [57], and another, called *outer consistency*, was introduced in [2].

Outer consistency As in the discussion of consistency, let D and D' be any two sets of distances, where $D' \neq D$. Then D' is called *outer-consistent* with the pair $(D, F(\mathcal{S}, D))$, if $D'(i,j) \geq D(i,j)$ for every pair of elements (i,j) that are in *different* clusters of $F(\mathcal{S}, D)$; and $D'(i,j) = D(i,j)$ for every pair of elements (i,j) that are in the *same* cluster of $F(\mathcal{S}, D)$.

A clustering algorithm, F, is called *outer-consistent* if for any two sets of distances, D and D', when D' is outer-consistent with the pair $D, F(\mathcal{S}, D))$, the clustering $F(\mathcal{S}, D')$ is the *same* as the clustering $F(\mathcal{S}, D)$.

That is, starting from a clustering $F(\mathcal{S}, D)$, if the distances between pairs of elements inside the same cluster stay the same, and the distances between pairs of elements in different clusters stay the same or increase, then the clustering does *not* change.

But, outer consistency is less restrictive than consistency, because it doesn't say what happens when distances between pairs of elements inside the same cluster *change*. An outer-consistent algorithm can produce *different* clusterings when the distances inside the same cluster *decrease* (or even increase) and the distances between clusters stay the same or increase. That outcome is not allowed by an algorithm that obeys Kleinberg's original consistency axiom.

The condition of outer consistency is also somewhat more justified than the condition of consistency, because there are well-known clustering algorithms that are outer-consistent (see [2]).

> **Review question:** Explain why every consistent clustering algorithm is also outer-consistent. Is it then correct to say that outer consistency is a weaker assumption than consistency?

Since outer consistency allows outcomes that are not allowed by consistency, it is *conceivable* that impossibility results involving consistency would *not* hold when only outer consistency is required.

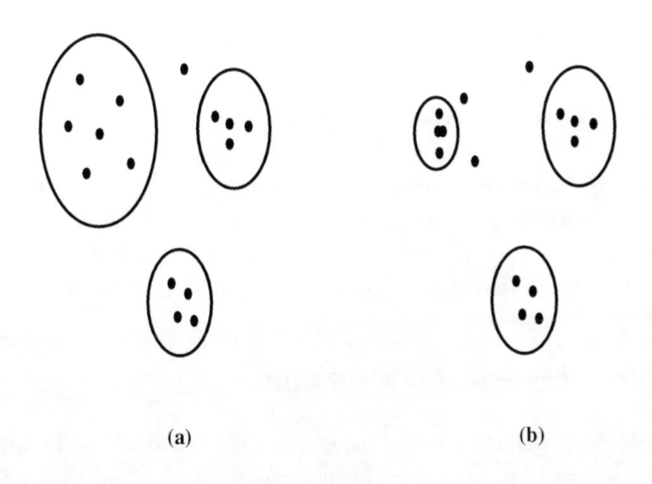

Figure 5.5 Panel (a) is the same clustering shown in panel (a) of Figure 5.1. In panel (b), the points in the upper-left cluster have been moved so that the distance between any two points in the cluster has either decreased or remains unchanged. Every other point in Panel (b) is unchanged from its placement in Panel (a). Visually, it now seems more natural to refine (split) the upper-left cluster into three clusters, two of which contain only a single point. The refined clustering is shown in panel (b).

However In the proof we presented for Theorem 5.2.1, every element in $\mathcal{S}$ is in its own cluster in $C_0 = F(\mathcal{S}, D)$, so D' is outer consistent with the pair $(D, F(\mathcal{S}, D))$. Therefore, the same proof also establishes:

Theorem 5.2.2 *There is no clustering algorithm that simultaneously obeys the axioms of scale invariance, richness, and outer consistency.*

Refinement Suppose $F(\mathcal{S}, D)$ and $F(\mathcal{S}, D')$ are two different clusterings of a set $\mathcal{S}$, for distance functions D and D', respectively, and recall that a cluster is a subset of elements of $\mathcal{S}$. If every cluster in $F(\mathcal{S}, D')$ is a subset of a cluster in $F(\mathcal{S}, D)$, then clustering $F(\mathcal{S}, D')$ is called a *refinement* of clustering $F(\mathcal{S}, D)$. That is, clustering $F(\mathcal{S}, D')$ may differ from clustering $F(\mathcal{S}, D)$ by *splitting* some clusters in $F(\mathcal{S}, D)$ into two or more clusters. Note that a set is considered a subset of itself, so the definition of refinement also allows two clusterings to be identical. See Figure 5.5.

Refinement consistency When should refinements of clusterings be allowed by a clustering algorithm?

When we decrease, or maintain, distances within clusters and increase or maintain distances between clusters, the consistency axiom requires that a clustering algorithm keep the *same* clusterings. But with such changes in distances, it may be more meaningful to allow the second clustering to be a *refinement* of the first one. This leads to a change in the axiom of consistency to:

The axiom of refinement consistency A clustering algorithm, F, is called *refinement-consistent* if for any two sets of distances, D and D', when D' is consistent with the pair $(D, F(\mathcal{S}, D))$, then the clustering $F(\mathcal{S}, D')$ must be a *refinement* of the clustering $F(\mathcal{S}, D)$. Kleinberg, in [57], claims:

Theorem 5.2.3 *There is no clustering algorithm that simultaneously obeys the axioms of scale invariance, richness, and refinement consistency.*

It is brittle Note that if a clustering algorithm F does *not* allow the anti social clustering, where each element in $\mathcal{S}$ is in its own cluster, then the proofs given for Theorems 5.2.1 and 5.2.2 break down. The same break down is stated in [57] to be true for Theorem 5.2.3. In fact, in [57] it is claimed that there is an algorithm that *does* obey the axioms of *scale invariance*, *richness*, and *refinement consistency*, *if* the algorithm is explicitly not allowed to produce the anti social clustering. Because of that, Kleinberg calls Theorem 5.2.3 true, but "brittle".

A more comprehensive critique of Kleinberg's axioms and impossibility theorem appear in [3]. That paper also proposes an alternative set of axioms that are similar in spirit to Kleinberg's, but allow clustering algorithms that *do* obey the alternative *scale invariance*, *richness*, and *consistency* axioms.

> **Review question** In the proof we presented for Theorem 5.2.1, C_0 is the antisocial clustering. So, if the antisocial clustering is not allowed, that proof does not work. Does this prove Kleinberg's claim that Theorem 5.2.1 does not hold if an algorithm is never allowed to produce the antisocial clustering? Explain.

5.3 Take-Home Lessons

The story of Kleinberg's impossibility theorems, and the critiques of them, illustrates an important point. As was true for Bell's theorems and for Arrow's theorems, when trying to prove that some task or phenomena are impossible, we must first detail a precise *model* of what we mean by that task or phenomena.[3] Only after a model is fully specified do we have the possibility of

[3] When we discuss Gödel's theorems, we will see this issue on steroids, because the gap there between what the precise model says and the imprecise way that people sometimes state Gödel's theorems is huge.

proving impossibility. But, the proof of impossibility in terms of the chosen model, while suggestive of possible impossibility outside of the model, is only assured for that precise model. One can hope (or fear) that the impossibility result extends beyond the precise model, but it is not always so.

In the case of Bell's theorems, impossibility does extend beyond Bell's original model based on EPR experiments. For example, Bell's impossibility results were later extended (and strengthened) to GHZ experiments. Arrow's impossibility theorems also extend beyond his original voting model, for example, in the GS theorem where the voting mechanism determines a *single* winner, rather than a full rank ordering of the candidates. But, Kleinberg's original result, as suggestive and impressive as it is, seems to have a more limited scope beyond the precise model used for that result. This is not a failing, but just part of the ebb-and-flow of applied research. This point is reflected in the following quote from Arthur C. Clarke:

> The only way of discovering the limits of the possible is to venture a little way past them into the impossible.

5.3.1 Proof vs Practice

There is another subtle point I want to make about the value of impossibility proofs illustrated by the brittleness of Theorem 5.2.2. As noted earlier, there are clustering algorithms that obey the axioms of *scale invariance*, *richness*, and *refinement consistency*, *provided* that those algorithms *forbid* the anti social clustering, where each element is in its own cluster. And since the anti social clustering is not likely to be informative in real applications, this is a pretty good outcome for practical clustering.

But, the *Gold Standard* of algorithm design is a mathematical *proof* of whatever properties are claimed or desired for an algorithm. The goal of mathematically oriented algorithm designers is to create an algorithm whose properties are *proved*, not just observed in practice.[4]

Now imagine such an algorithm designer, before Kleinberg's paper, who is trying to design a clustering algorithm that *always* obeys the above three axioms and who wants to *prove* that their proposed algorithm obeys the axioms. That algorithm designer would always fail, no matter what algorithm they devised or how well it performed in practice. Theorem 5.2.2 guarantees that. But, before Kleinberg's paper, they wouldn't know this for sure, or know

[4] The classic joke (but actually making a serious point about the importance of theory) is the following question: "Sure it works in practice, but does it work in theory?"

why they were failing. So, they might just keep trying to find that elusive algorithm.

After the proof of Theorem 5.2.2, the designer, while perhaps disappointed, would now understand the futility of their efforts and the need to restrict their focus. And that refocusing can be aided by seeing the way that the proof of Theorem 5.2.2 fails after a small change in the model.

The take-home lesson here is that while Kleinberg's theorems may have limited impact on the practice of clustering, they have real impact on the logic and mathematical investigation of clustering.

5.4 Exercises

1. Generally, the antisocial clustering (where every element is in its own cluster) is not likely to be informative or useful for real clustering problems. But that is not always true. Give a pictorial example where the antisocial clustering might be highly informative.
2. Theorem 5.2.1 can be proved by letting C_0 be the clustering where all of the elements in $\mathcal{S}$ are in *one, single* cluster, instead of where C_0 is the antisocial clustering. That proof is a modification of the proof presented for Theorem 5.2.1. Find this modified proof.
3. *Inner consistency* is a property of a clustering method that is symmetric to outer consistency. It is defined as:

   For two sets of distances $D' \neq D$, D' is *inner-consistent* with $(D, F(\mathcal{S}, D))$, if $D'(i,j) = D(i,j)$ for every pair of elements (i,j) that are in *different* clusters of $F(\mathcal{S}, D)$ and $D'(i,j) \leq (D(i,j)$ for every pair of elements (i,j) that are in the *same* cluster of $F(\mathcal{S}, D)$.

   A clustering algorithm, F, is called *inner-consistent* if for any two sets of distances, D and D', when D' is inner-consistent with $(D, F(\mathcal{S}, D))$, the clustering $F(\mathcal{S}, D')$ is the same as $F(\mathcal{S}, D)$.

   Prove that there is no clustering algorithm that obeys the axioms of *scale invariance*, *richness*, and *inner consistency*.
4. If an algorithm is both outer-consistent and inner-consistent, is it necessarily consistent? If no, give an example; and if yes, prove it.
5. The axioms of *consistency*, *outer consistency*, *refinement consistency*, and *inner consistency* all involve increasing or maintaining distances between clusters, and decreasing or maintaining distances inside clusters. But, changes in distances might also decrease distances between clusters, and/or increase distances inside clusters.

   Give examples (in dot figures, as in Figure 5.1) where such changes might be reasonable.

Do any of the impossibility results discussed in this chapter apply when such changes are allowed? At first it seems that they might, because if distances between clusters increase when moving from distance D to distance D', then distances decrease when moving from D' to D. Is this line of reasoning helpful?

6. Is there any relationship between the axioms of outer consistency and refinement consistency? That is, does one axiom imply the other?

6
A Gödel-ish Impossibility and Incompleteness Theorem

A Rigorous, Streamlined Proof of (a simpler variant of) Gödel's First Incompleteness Theorem, Requiring Minimal Background[1]

6.1 Introduction

> In the first quarter of the twentieth century there were some mathematical systems in existence that were so comprehensive that is was generally assumed that every mathematical statement could either be proved or disproved within the system. In 1931 Gödel astounded the entire mathematical world by showing that this was not the case. For *each of the mathematical systems in question*, there must always be mathematical statements that can neither be proved nor disproved within the system. [101] (italics added)

Gödel's famous incompleteness theorems (there are two of them, called *First* and *Second*), proved in the 1930s, concern the limitations of formal, axiom-based, logical systems to express and derive statements about integer arithmetic,[2] and more generally, about mathematics. When people refer to

[1] Other than common things, you only need to know what an *integer* is, and what a *function* is; and you need to know that a computer program is a series of *textual* statements in some language: "A reader who is not familiar with any programming language need only think of it as a language for expressing [textual] instructions on how to proceed in a mechanical step-by-step computation" [39].

The material in this chapter was developed to be presented in the first few lectures of my course offering "The Theory of Computation", which is a sophomore-level college course. You do need a few hours to absorb the material, and you need to focus – the material is concrete and requires almost no background, but it is not trivial. Have pen and paper ready, and use them.

[2] Integer arithmetic must include, at least, the operations of addition and multiplication of nonnegative integers, that is, nonnegative whole numbers.

"Gödel's incompleteness theorem", they generally mean Gödel's *First* incompleteness theorem.

Gödel's theorems are well known to logicians, philosophers, and mathematicians, and Gödel is widely considered to be the greatest logician since Aristotle. Gödel's theorems (or at least their existence and rough statements) became widely known to the general public through two books: *Gödel's Proof* by Nagel and Newman [77] in 1958 and the best-selling book *Gödel, Escher and Bach* by Douglas Hofstadter [51] in 1979.

Since then, several general-public books and many articles on Gödel and Gödel's proofs have been published – the most recent book in 2021 [20].

A Popular Formulation of the First Incompleteness Theorem

> The incompleteness theorem is often popularly formulated as saying that for systems Π to which the [Gödel] theorem applies, there is some *true* statement in the language of Π that is undecidable [neither provable nor disprovable] in Π. [39][3]

Gödel flavors: semantic and syntactic Gödel's First incompleteness theorem comes in two flavors: the *semantic* version and the *syntactic* version. Sharpening the popular formulation of incompleteness stated above, the *headline* summary of the *semantic* version of Gödel's First incompleteness theorem is:

> In any axiom-based, logical system for deriving statements about arithmetic, if a few basic facts about arithmetic[4] can be derived in the system, and *only* true statements can be derived in the system, then there must be true statements about arithmetic that *cannot* be derived in the system.[5]

[3] The original symbol *S* has been replaced by Π for consistency with notation in the rest of the book.

[4] We are leaving this vague, but it turns out that very few facts are needed.

[5] In the chapters of this book that concern mathematical logic (Chapters 6, 7, 8, and 9), I use the terms "derive" and "derivation" for what the system can do, when the words "prove" and "proof" might be more typically used. This is because the words "prove" and "proof" have the popular connotation of establishing something that is necessarily *true*. But much of mathematical logic, and certainly much of incompleteness theory, concerns the ability of a formal system to reach a statement (i.e., derive it) from an initial set of axioms by applying logical rules, where it is not always the case that what the system reaches is "true". Quoting David Deutsch on the podcast ToKast Ep. 82:

> When we say that something is provable we mean that it is provable from certain axioms. Gödel's theorem is about provability, which means provability from axioms ... Gödel's theorem is not about establishing truths, it is about deriving things from other things.

Of course, the ideal result is when "derivation" and "proof" go hand in hand, but that is not always the case. Quoting from [99]: "... we want formal derivations which will be proofs in the intuitive sense if we can take it that the axioms are indeed secure truths."

The *headline* summary of the *syntactic* version of Gödel's First incompleteness theorem is:

> In any axiom-based logical system for arithmetic, if a few basic facts about arithmetic can be derived in the system, and if the logical system can never derive a *contradiction* (i.e., can never derive both a statement and its negation), then there will be a statement about arithmetic, such that *neither* it nor its *negation* can be derived in the system.

More loosely, if such a logical system is free of contradictions, then there will be statements that can neither be proved nor disproved in that system – those statements are said to be *undecidable* in the system, and the system is called *incomplete*. And this implies that it is *impossible* to find a "single formal system within which all mathematical problems are solvable, even if we restrict ourselves to arithmetical problems" [39]. This impossibility was not welcome news. Quoting Chaitin [23]:

> At the time of its discovery, Kurt Gödel's incompleteness theorem was a great shock and caused much uncertainty and depression among mathematicians sensitive to foundational issues, since it seemed to pull the rug out from under mathematical certainty, objectivity, and rigor.

Gödel didn't even need or want truth Notice that the semantic version of Gödel's theorem concerns *truth*, while the syntactic version only concerns *consistency*, that is, avoidance of contradiction. That distinction was very important to Gödel. As stated in [99]:

> Gödel was writing at a time when, for various reasons ... the very idea of truth-in-mathematics was under some suspicion. It was therefore *extremely* important to Gödel that he could show you don't need to employ any semantic notions to get an incompleteness result.

6.1.1 In This Chapter

In this chapter, we discuss a *variant* of the *semantic* version of Gödel's *First* incompleteness theorem, giving full, rigorous proofs of results that establish the popular formulation of incompleteness, stated above. However, what we rigorously prove will be technically weaker, in several ways, than what Gödel actually proved in his First incompleteness theorem.[6] We will discuss this more fully in the next two chapters.

For conclusions reached using every-day mathematics *outside* of a formal axiomatic system, we will still use the words "prove" and "proof".

[6] In most mathematics, philosophical, and computer science literature, the term "Gödel's Theorem" has been applied fairly liberally. Once the full technical content of Gödel's

What we do in this chapter follows the modern, post-Gödel approach to incompleteness that exploits *non-computability*, using ideas due mostly to Alan Turing, and not yet developed when Gödel proved his incompleteness theorems. We fully *prove* a Gödel-*ish* incompleteness result and introduce many (but not all) of the key techniques used in other proofs of incompleteness.

We will also discuss, but without proofs, the *syntactic* version of Gödel's First incompleteness theorem. In subsequent chapters, we will discuss incompleteness and non-computability theorems that also imply Gödel-ish theorems. All told, we will fully prove *four* incompleteness theorems that each align with the popular formulation of incompleteness, stated at the start of this chapter. However, none of those proofs will *exactly* prove what has become known as "Gödel's First incompleteness theorem".[7] Proving Gödel-ish incompleteness theorems is the common, modern approach to incompleteness.[8]

Later, in Chapter 9, we will discuss Gödel's *Second* incompleteness (or inconsistency) theorem and introduce the key ideas in its proof (but we will not actually prove Gödel's Second theorem).

theorems and their philosophical import were understood, people proved other, sometimes technically weaker, incompleteness and undecidability results (some unrelated to logic and some unrelated to arithmetic) that have much of the same philosophical import as Gödel's First theorem. Often, despite the differences, the authors still called their result "Gödel's Theorem". For example, George Boolos published the paper titled "A New Proof of Gödel's Incompleteness Theorem" [17], but in the paper he states "... we shall give an easy new proof of the Gödel Incompleteness Theorem in the following form: 'There is no algorithm whose output contains all true statements of arithmetic and no false ones'." That focus on algorithms makes no explicit connection to logic, and hence is certainly different from what Gödel proved. Following that practice, we could have entitled this chapter "A Gödel Theorem", but chose "Gödel-ish" instead for greater clarity.

[7] In fact, Gödel didn't either.

[8] In comparing the more modern approach, which is technically much simpler than Gödel's original proof, it is helpful to realize that Gödel had a driving goal of great importance at the time, which today is mostly of historical interest. That goal was to show that the project of Russell and Whitehead to axiomatize all of mathematics (starting with arithmetic) was fatally flawed. In particular, the logical system in *Principia Mathematica (PM)* was *incomplete*, that is, there were statements about arithmetic (number theory) where neither the statement nor its negation could be derived in PM. Incompleteness also appeared in other logical systems. Much of the grungy detail of Gödel's proof involves what he did to connect *high-level* thinking about incompleteness generally to the *low-level* detail of PM, and related axiomatic systems. This is still of importance to logicians, but today the phenomena of incompleteness and non-computability (also called *undecidability*) are more important in applications *other* than the study of axiom-based formal systems. We will discuss the main ones in Chapters 7 and 8. Hence, the modern approach to incompleteness is technically simpler, and it focuses its energies on why and how incompleteness is currently important.

And yes, this is a defense against the objection that I have not included in this book a "full, real" proof of Gödel's First incompleteness theorem. But in Chapter 9, we will fully prove an *abstract* version of Gödel's First incompleteness theorem, that does go back to Gödel.

6.2 The First Key Idea: There are Non-computable Functions

We assume that the reader is familiar with the concept of a function. A *binary* function is a function from the positive integers to {0, 1}. That is, f is a binary function if for any positive integer x, the value of $f(x)$ is either 0 or 1. For example, the *parity* function f that indicates whether a positive integer is *even* or *odd* is a *binary* function. To be more precise, $f(x) = 0$ if and only if the value of x is an even positive integer, 2,4,6,8 ..., and $f(x) = 1$ if and only if the value of x is an odd positive integer, 1,3,5,7, ...

We use K to denote the set of *all* binary functions from the positive integers to {0, 1}. Note that the number of functions in K is infinite.

A function f in K is called *computable* if there is a computer program that can compute function f.

That is, given *any* positive integer x, the program to compute $f(x)$ finishes in finite time and correctly outputs the value $f(x)$.[9] It is assumed that any computer program must be of finite size, although there is no *a priori* limit on the size of a program.

Certainly, for example, we can write a computer program to determine if a positive integer is even or odd, that is, to determine the *parity* of a positive integer.

We use C to denote the set of functions in K that are computable. So, to exercise this notation, we can say that the parity function is in set C.

Note that the number of functions in C is *infinite*. For example, the function $f(7) = 1$ and $f(x) = 0$ for every integer $x \neq 7$ is a computable function, and we can create a similar computable function for any positive integer, in place of 7. So, since there are an infinite number of positive integers, there are an infinite number of computable functions. Of course, there are many computable functions that are much more interesting and important than the example above.

However, *not* all binary functions are computable, as we will prove in the next theorem. This theorem exposes the heart and soul of incompleteness and undecidability, both of the Gödel type, discussed in this chapter, and of the Turing type, discussed in Chapter 7.

[9] We don't need to be more specific, but if you insist on being super-concrete, we could say that the program is written in Python 3.7.3, and that it is executed on a MacBook Pro (with a silver case) running Snow Leopard OSX.

Theorem 6.2.1 *There are functions in K that are* not *computable. That is,* $C \subset K$.[10]

Before we prove Theorem 6.2.1, we need a definition and concept. Choose a computer language and consider all the programs in that language that compute a binary function. Each line in a program has some end-of-line symbol, so we can concatenate (i.e., join) the lines of the program together into a single long *string*.[11] Therefore, we can think of the program as a single string written using some finite alphabet. Such a string is called a *string-encoding* of the program it comes from.

Proof of Theorem 6.2.1

Consider a list, L_S, of all strings made from the alphabet of the chosen computer language. We can order the strings in L_S by their lengths (shortest first), and using lexicographic order (i.e., the way strings are ordered in a dictionary) for strings of the same length. So, for any positive integer, i, all strings of length i appear in the list before any strings of length $i+1$; and the strings of length i are ordered lexicographically. Such a list L_S conceptually exists although we can't actually create it since it has an infinite number of entries.

For example, suppose (to make the illustration simple) that the alphabet of the computer language just consists of three characters, a, b, and c. Then, the ordered list L_S of all the strings made from this alphabet begins with:

a, *b*, *c*, *aa*, *ab*, *ac*, *ba*, *bb*, *bc*, *ca*, *cb*, *cc*, *aaa*, *aab*, *aac*, *aba*, *abb*, *abc*,
aca, *acb*, *acc*, *baa*, *bab*,
bac, *bba*, *bbb*, *bbc*, *bca*, *bcb*, *bcc*, *caa*, *cab*, *cac*, *cba*, *cbb*, *cbc*, *cca*, *ccb*,
ccc, *aaaa*, *aaab*, *aaac*...

Review question What are the next six strings in L_S after *aaac*?

Of course, none of the above strings is a computer program (in any computer language that I know), but for real computer languages with a larger alphabet, when the strings in L_S get long enough, *some* of those strings must be string-encodings of computer programs in the chosen computer language.

[10] The symbol "$\subset$" means "is a subset of, but not equal to". So $C \subset K$ means that C is a subset of K, but not equal to K. So C is a *strict* subset of K. Although not used here, we note that the symbol "$\subseteq$" means that the subsets might be identical.

[11] In some languages (such as Python) spaces are significant, so the single string must include any spaces (or tab symbols) in the original program.

List L_C If we (conceptually) remove every string from L_S that is *not* a string-encoding of a program in the chosen computer language, what remains is an ordering of the string-encodings representing programs in that language. Some of those programs compute binary functions (i.e., functions in K), but most do not. Then, if we conceptually remove from L_S any string-encoding of a program that does not compute a binary function, what remains is a list of all the string-encodings of programs that compute functions in set C. That list is called L_C.

So L_C is a list of all string-encodings of programs that compute binary functions. List L_C is ordered by the length of the strings where ties are broken lexicographically, as described above. So each *program* that computes a function in C has a well-defined *position* in L_C, and so is associated with a positive integer that specifies its position in L_C.[12]

A function in C might be computed by different computer programs, so a function f in C might be represented in L_C more than once, but that will not matter. Each program has a well-defined position in L_C.

Recapping: List L_C is an ordered list of string-encodings of programs computing all the functions in C, where a function can be represented in L_C more than once.

We use f_i to denote the binary function that is computed by the ith program in L_C. (Remember that list L_C is only conceptual; we don't actually build it – we just have to *imagine* it for the sake of the proof.)

Table T Next, consider a table T with one column for each positive integer, and one row for each program in L_C; and associate the function f_i with row i of T. Then set the value of cell $T(i, x)$ to $f_i(x)$. See Table 6.1. Since L_C is only conceptual, T is also only conceptual.[13]

The key function $\bar{f}$ Next, define a new function, $\bar{f}$, from the positive integers to {0, 1} as:

$$\bar{f}(i) = 1 - f_i(i).$$

[12] We note that an infinite set of objects where there is a conceptual way to assign each object a distinct positive integer, and each positive integer is assigned to one object in the set, is said to be *countably infinite*. So the set of programs in L_C is countably infinite. We will use this fact in Chapter 7.

[13] Roger Penrose, in [83], defines a similar table (he calls an *array*) and then states: "I am not asking [asserting] that we have actually *calculated* this array, say by some algorithm ... We are just supposed to *imagine* that the *true* list has somehow been laid out before us, perhaps by God!" (the italics are original).

Table 6.1 *The conceptual Table T is an enumeration of all computable binary functions (from positive integers to* {0, 1}*), and their values at all of the positive integers.*

	1	2	3	4	5	.	x	.	i	...
f_1	1	1	0	0	0		$f_1(x)$			
f_2	0	0	1	0	0		$f_2(x)$			
f_3	1	1	0	0	1		$f_3(x)$			
f_4	0	0	1	1	0		$f_4(x)$			
f_5	0	1	0	0	0		$f_5(x)$			
.						.				
.							.			
.								.		
f_i									$f_i(i)$	
⋮										⋱

For example, based on the functions in Table 6.1, $\overline{f}(1) = 0, \overline{f}(2) = 1, \overline{f}(3) = 1, \overline{f}(4) = 0$, and $\overline{f}(5) = 1$.

Note that in the definition of $\overline{f}(i)$, the same integer i is used both to identify the function f_i in C and as the input value to f_i and to $\overline{f}$. Hence, the values for $\overline{f}$ are determined from the values along the main *diagonal* of table T. Note also that function $\overline{f}$ changes 0 to 1 and changes 1 to 0. So, the values of function $\overline{f}$ are the *opposite* of the values along the main diagonal of Table T. Clearly, function $\overline{f}$ is a binary function, that is, it is in K.

Now we ask and answer Is $\overline{f}$ a computable function? That is, is it in C?

The answer is *no* for the following reason. If $\overline{f}$ were in C, then there must be some row i^* in T such that $\overline{f}(x) = f_{i^*}(x)$ for any positive integer x. For example, maybe i^* is 57. Is that possible? No, because $\overline{f}(57) = 1 - f_{57}(57) \neq f_{57}(57)$, so $\overline{f}$ can't be f_{57}.

More generally, $\overline{f}(i^*) = 1 - f_{i^*}(i^*)$, so $\overline{f}$ and f_{i^*} differ *at least* for one input value, namely i^*. So function $\overline{f}$ cannot be identical to function f_{i^*}. Hence, there is *no* row in T corresponding to the binary function $\overline{f}$. But, table T is defined to have at least one row for every function in C. So, function $\overline{f}$ is *not* in C, and $\overline{f}$ is not computable – it is in set K, but not in C. And that completes the proof of Theorem 6.2.1. ■

The above proof is an example of a *proof by diagonalization* (a technique first developed by Georg Cantor), since the changes in T are made along the main diagonal of table T. Such proofs are widely used in obtaining non-computability and incompleteness results.[14]

For emphasis Again, the list L_C and the table T are only conceptual. We don't know how to construct L_C and T, but that does not matter. We can conceive of them in order to *define* function $\overline{f}$.

Again, and again: The *set* of computer programs (in your selected language) can (in principle) be ordered by length (and lexicographically for programs of the same length), and the set of computer programs that compute binary functions is a subset of all programs, so there is also an ordering of computer programs that compute the binary functions. (Thus, the set of computer programs in L_C is *countably* infinite.) Each computable binary function is computed by one or more of these programs, and we can associate each computable binary function f with the *first* program in list L_C that computes f. So, this tells us that there *is* a well-defined ordering of the computable binary functions, even if we don't know how to construct it. And that means that $\overline{f}$ is a real, well-defined binary function, even though the way it is defined is very abstract and very different from the way that functions are typically defined.[15]

6.3 What Is a Formal Proof System?

How do we connect Theorem 6.2.1, which is about *functions*, to (our version of) Gödel's First incompleteness theorem, which is about *logical systems*? We first must define a logical system, also called a *formal proof system* or just a *formal system*.

A *formal (proof) system*, Π, has three components:

1) A *formal language* consisting of a finite *alphabet* and a finite subset of words and phrases written in that alphabet that can be used in forming (or writing) *statements*.

[14] Diagonalization, such as in this proof, is a much less mysterious form of *self-reference* than is often associated with Gödel and incompleteness. We will discuss this further in Chapter 9.

[15] I am making this point super explicitly, because some feedback I received from an earlier version of this chapter objected to the use of $\overline{f}$ due to the tortured, non-constructive way it is defined.

2) A finite list of *axioms* (statements that we take as being true).[16] The axioms are written in the formal language of the system.
3) A finite list of *logic rules*, also called *inference* rules, that can be applied to transform, in a precise *mechanical* way, one statement into another.[17]

For example, the alphabet might be the standard ASCII alphabet with 256 symbols. A first axiom might be the statement "for any integer x, $x + 1 > x$." A second axiom might be "for any integers x and y, $x + y$ is an integer." A logic rule might be "for any three integers, x, y, z, if $x > y$ and $y > z$ then $x > z$." (Call this rule the "transitivity rule".) A similar logic rule might be the famous *modus ponens*, which says:

> If statement A has been established, and "statement A implies statement B" has been established, then we can conclude that statement B is also established.

Another logic rule might be "for any two functions represented by f and f', if $f(x) = f'(x)$ for every value of x, then you can replace f with f' in any statement containing f." (Call this the "Replacement rule".)

The finite set of English words and phrases might include the phrase "for any integer". Of course, there will typically be more axioms, logic rules, and allowed words and phrases than in this example.

6.3.1 What Is a Formal Derivation?

A *formal derivation* of a statement, S, in the formal system Π, is a series of statements that begin with some axioms of Π, and then successively use logic rules in Π to obtain (derive) the statement S.

For example, S might be the statement: "for any integer x, $x + 1 + 1 > x$." Using the two axioms stated above, *modus ponens*, the replacement rule and the transitivity rule, a formal derivation of statement S in Π is:

1. For any integer x, $x + 1 > x$ (by the first axiom).
2. x is an integer, and 1 is an integer, so $x + 1$ is an integer (by the second axiom).

[16] Actually, in the *syntactic* version of Gödel's theorem there is no concept of "truth", so a set of axioms can just be a starting point for formal derivations. We will not explore that possibility now, both in order to keep things simpler and because in typical mathematical use, axioms are statements that are accepted as being true. There are also formal systems with an *infinite* number of axioms (that must obey certain properties), but we will not discuss those.

[17] I don't know why these aren't called "deduction" rules, since they actually correspond to deductions rather than inferences. But, I didn't create the terminology.

3. Define y to be equal to $x + 1$ and use the replacement rule on the first statement, resulting in: $y > x$.
4. y is an integer since $x + 1$ is an integer (more formally, this comes from replacing $x + 1$ by y in the second part of the second statement above).
5. $y + 1 > y$ (by the first axiom, since y is an integer).
6. Using the replacement rule, replace y with $x + 1$, resulting in: $x + 1 + 1 > x + 1$.
7. Use the first axiom (i.e., $x + 1 > x$) and the transitivity rule to conclude that $x + 1 + 1 > x$, which is statement S.

The finite subset of English used in this formal derivation includes the words and phrases "Define", "to be equal", "Using the replacement rule", "We conclude", and so forth. These would be part of the finite subset of English that is part of the definition of Π. Each phrase used must have a clear and precise meaning in Π, so that each statement in a formal derivation, other than an axiom, follows in a *mechanical* way from the preceding statements, by the application of one logic rule. Each step of the formal derivation must also state which logic rule is employed in the step.

It's tedious, but precise Formal derivations are very tedious, but each step is explicitly justified. We humans don't want to write derivations or proofs this way, but computers can do it, as we will explain below.

6.3.2 Mechanical Generation and Checking of Formal Derivations

We now make four key points about formal derivations:

1. Similar to the way we argued that list L_C exists, the first key point is that it is easy to write a computer program, P, that can begin generating, in order of the length of the string, every string s that can be written in the alphabet of Π, augmented by the allowed words and phrases of the formal system Π. Program P will never stop because there is no bound on the length of the strings. Most of the strings will not be formal derivations of anything, but for any finite-length string s using the alphabet of Π, P will eventually (and in *finite* time) generate s.

2. The second key point is that any formal derivation, being a series of statements, is just a *string* formed from the alphabet and the allowed words and phrases of the formal system Π. This is analogous to the way we treated

computer programs as strings. Hence, if s is a formal derivation, program P (generating strings) will eventually (and in finite time) generate it.

3. The third key point about a formal system is that we can create a computer program P' that knows the alphabet, the allowed phrases, the axioms, and the inference rules, so that P' can precisely interpret the effect of each statement in a formal derivation. That is, P' can *mechanically* check whether each line is an axiom, or whether it follows from the previous statement(s) by an application of some inference rule.

Therefore, given a statement S and a string s that might be a formal derivation of S, program P' can check (in a purely mechanical way, and in finite time) whether string s is a formal derivation of statement S in Π.

4. The fourth key point is that for any statement S, after program P generates a string s, program P' can check (in finite time) whether s is a formal derivation of statement S in Π, before P generates the next string. Hence, if there is a formal derivation s in Π of statement S, then s will be generated and recognized in finite time by interleaving the execution of programs P and P'.

Note that most of the strings that P generates will be garbage, and most of the strings that are not garbage will not be formal derivations of statement S. But if string s is a formal derivation of statement S in Π, then in finite time, program P *will* generate s, and program P' *will* recognize that s is a formal derivation in Π of statement S.

6.4 Back to Gödel and Incompleteness

How do we connect all this to Gödel's First incompleteness theorem? We will focus on function $\overline{f}$ and show that in any formal system, Π, that can only derive true statements, there must be some true statement about $\overline{f}$ that *cannot* be derived in Π.

Before proceeding, recall that the binary function $\overline{f}$ is well defined, that is, there is a value $\overline{f}(x)$ for every positive integer x; and for any specific x, $\overline{f}(x)$ is either 0 or 1. Recall also that $\overline{f}$ is *not* a computable function – there is no computer program that can compute $\overline{f}(x)$ for every positive integer x.

We call a statement an $\overline{f}$-*statement* if it is either:

$\overline{f}(x)$ `is 1`
or:
$\overline{f}(x)$ `is 0`

for some positive integer x.

Note that every $\bar{f}$-statement is a statement about a *specific* integer. For example, the statement "$\bar{f}(57)$ `is` 1" is an $\bar{f}$-statement, where x has the value 57. Since, for any positive integer x, $\bar{f}(x)$ has only two possible values, we refer to the first statement as $Sf(x)$ and the second statement as $\neg Sf(x)$.

6.4.1 What Is Truth?

For any positive integer x, we say that the $\bar{f}$-statement "$Sf(x)$" is *true*, and statement "$\neg Sf(x)$" is *false*, if in fact $\bar{f}(x) = 1$. Similarly, we say an $\bar{f}$-statement "$\neg Sf(x)$" is true, and "$Sf(x)$" is false if in fact $\bar{f}(x) = 0$.[18]

Clearly, for any positive integer x, one of the statements "$Sf(x)$" or "$\neg Sf(x)$" is true and the other is false. In this context, truth and falsity are simple, well-defined concepts, although "truth" is not so simple to define in general.[19] It is a natural and desirable property of a formal system Π (where truth has a clear definition) that no false statement can be derived in Π. In fact, this property has a name:

A formal system Π, in which truth has been defined, is called *sound* if every statement that can be derived in Π is true. That is, if Π is sound, then it is not possible to derive a false statement in Π.

Note that the concept of "soundness" of a formal system only applies when truth has been defined for that system (as we have done in the $\bar{f}$ system). But, a formal system need not have a defined truth – it might only be about manipulating meaningless symbols. As we will see, one (well, Gödel in particular) can still deduce properties of such "meaningless" systems, where there is no concept of truth.

$\bar{f}$-Expressible systems We define a formal system Π as $\bar{f}$*-expressible* if it can form (write) any $\bar{f}$-statement in Π. Note that the words "form" and "write" do

[18] At first exposure, these definitions seem a bit empty, or circular. But, keep in mind that we are defining truth and falsity of *statements* about function $\bar{f}$ in relation to the *reality* of function $\bar{f}$. And consider the following: Most discussions of truth in logic begin with the following sentence:

" 'snow is white' is true because snow is white".

Now that really seems circular and empty. But the first phrase of the sentence is in single quotes, meaning that the subject of the sentence is a *statement* – the subject is *not* snow. So, the sentence is asserting that the statement "snow is white" is true because in fact, snow (that cold, fluffy form of water) is white (at least for the sake of this discussion). Now, doesn't that make you want to be a logician?

[19] In fact, there is a theorem called *Tarski's* theorem, related to Gödel's First theorem, which says that it is not always possible to define truth inside certain logical systems. I was going to include a discussion of Tarski's theorem in this book, but in truth, I don't really understand truth.

not mean "derive". The question of whether a statement can be derived in Π is at the heart of Gödel's incompleteness.

6.4.2 Finally, Our Variant of Gödel's Theorem

Theorem 6.4.1 *There is no $\overline{f}$-expressible formal proof system* Π *that can derive all true $\overline{f}$-statements and never derive a false $\overline{f}$-statement.*

Proof Suppose that there is an $\overline{f}$-expressible formal system Π such that:

(a) No *false* $\overline{f}$-statements can be derived in Π – only true $\overline{f}$-statements can be derived in Π.
(b) For any *true* $\overline{f}$-statement S, there is a formal derivation of S in Π.

Suppositions a and b together say that *all* true $\overline{f}$-statements, and *only* true $\overline{f}$-statements, can be derived in Π.

Now, since Π is $\overline{f}$-expressible, for any positive integer x, both statements $Sf(x)$ and $\neg Sf(x)$ can be formed in Π. But since exactly one of those two statements is true, suppositions a and b imply that there will be a formal derivation in Π of *exactly one* of those statements, in particular, the statement that is true. But this leads to a contradiction of the established fact that function $\overline{f}$ is not computable.

The key point In more detail, if the two suppositions, a and b, hold, then the following approach describes a computer program, P^*, that can, in finite time, correctly determine the value of $\overline{f}(x)$, for any positive integer x.

> **Program P^*:** Given a positive integer x, run program P to successively generate the strings in order of their lengths (using the finite alphabet and finite known words and phrases in Π), ordering same length strings lexicographically.
>
> After each string s is generated by program P, run program P' to see if s is a formal derivation of statement $Sf(x)$. If it is, statement $Sf(x)$ must be true, so P^* can correctly output that $\overline{f}(x) = 1$, and stop. But if program P' determines that s is not a formal derivation of $Sf(x)$, run P' again to see if s is a formal derivation of the statement $\neg Sf(x)$. If it is, P^* can correctly output that $\overline{f}(x) = 0$, and stop.
>
> If s is neither a formal derivation of $Sf(x)$ nor of $\neg Sf(x)$, continue running program P to generate the next string s.

P^* works The two suppositions (a and b) guarantee that for any positive integer x, the mechanical computer program, P^*, will have a Eureka moment in finite time, revealing the correct value of $\overline{f}(x)$. But, then $\overline{f}$ would be a *computable* function (computable by program P^*), contradicting the already established fact that $\overline{f}$ is *not* a computable function. This contradiction means that the suppositions a and b *cannot* both hold, proving Theorem 6.4.1. ■

6.5 Alternate Statements of (Our Variant of) Gödel's First Incompleteness Theorem

Incompleteness There are several equivalent conclusions we can draw from what we proved. One is that if an $\overline{f}$-expressible formal system Π has the property that only true statements can be derived in Π (i.e., supposition a), then there must be true statements that can be *formed* in Π but *not derived* in Π. That is, Π must be *incomplete*. More formally:

Theorem 6.5.1 *For any $\overline{f}$-expressible formal system Π in which only true $\overline{f}$-statements can be derived, there will be some true $\overline{f}$-statement that cannot be derived in Π.*

Note that we don't really care about the function $\overline{f}$ – it is just a device to prove the incompleteness of certain formal systems (the ones that are $\overline{f}$-expressible and never derive a false statement).[20]

Gödel's First theorem is stated in several different ways. Consistent with those alternate statements, we restate Theorem 6.5.1 as follows:

Theorem 6.5.2 *For any $\overline{f}$-expressible formal system Π in which no formal derivation of a false $\overline{f}$-statement is possible, there will be some $\overline{f}$-statement, S, such that* neither *S nor $\neg S$ can be derived in Π.*

Theorem 6.5.2 is immediate from Theorem 6.5.1. Consider a true $\overline{f}$-statement S that, by Theorem 6.5.1, cannot be derived in Π. Statement S is true, so $\neg S$ must be false, and since no false statement can be derived in Π, Theorem 6.5.2 follows.

Other common restatements Recall that a formal system Π is called *sound* if it is never possible to derive in Π any *false* statement.

A formal system Π is called *complete* if for every statement S that can be expressed in Π, either S or $\neg S$ can be *derived* in Π. Theorem 6.5.2 can then be restated as:

Theorem 6.5.3 *No formal $\overline{f}$-expressible system Π can be both sound and complete.*

[20] Also, the requirement that a formal system be $\overline{f}$-*expressive* is just to avoid being distracted by trivialities. We are not interested in formal systems that can't even form statements about $\overline{f}$ values.

And, although Theorem 6.5.3 is the main, fully proved, incompleteness theorem of this chapter, we also state the real Gödel theorem that corresponds to it.[21]

Theorem 6.5.4 *No formal system Π that can form statements about arithmetic, and derive certain very basic facts about arithmetic, can be both sound and complete.*

6.5.1 Consistency and Soundness

A formal system Π is called *consistent* if it is never possible to derive a statement S in Π and also derive its negation (denoted $\neg S$) in Π. Otherwise, Π is called *inconsistent.* A consistent system cannot derive *contradictory* statements, but an inconsistent system can.

Note that when truth has been defined for a system, the property of being *sound* is stronger than the property of being *consistent.* Every sound system must be consistent, but not every consistent system is sound.[22]

Note also that soundness is an absolutely desirable property of a formal system that is actually *about* some specific domain, for example, a subarea of mathematics, such as arithmetic or geometry, where we want to develop a formal system precisely because it can be used to derive *true* statements about that domain.

But amazingly, Gödel didn't need truth Formal systems can be studied without specifying a specific domain of application, and hence without defining "truth" or "meaning" in any domain. In that case, although we can't talk about a formal system being sound, we will still want it to be *consistent* (i.e., free of contradictions). And, now we can state (but not prove) what is called the *syntactic* version of Gödel's First incompleteness theorem:

Theorem 6.5.5 *No formal system Π that can form statements about arithmetic, and derive certain very basic facts about arithmetic, can be both consistent and complete.*

[21] To be clear, we have not proved this real Gödel theorem, but only Theorem 6.5.3.

[22] Consider the formal system, whose language is English, containing only one axiom: "unicorns exist", and two inference rules: *modus ponens* and "if unicorns exist, then all men are aardvarks". Convince yourself that this system can only derive the statements "unicorns exist" and "All men are aardvarks." The system cannot derive the negation of either of these two statements. Hence, the system is consistent. But, under the ordinary understanding of what is true about unicorns, men and aardvarks, the system is certainly not sound.

Notice that Theorem 6.5.5 differs from Theorem 6.5.4 in that the word "sound" has been replaced by the word "consistent". Theorem 6.5.5 similarly differs from Theorem 6.5.3. This is one way that Gödel's actual First incompleteness theorem is stronger than Theorem 6.5.3.[23] It is a theorem about *symbols* only – it does not need or involve any concept of *truth*. That was important to Gödel because "truth" is a somewhat squishy, vague concept. In this chapter, we avoided squishiness by focusing on a very precise, narrow domain, that is, that of $\overline{f}$-statements, where truth has an obvious meaning.

6.6 What Have We Learned, and What More Do We Want?

We have shown that there is no sound, formal system that can derive all true statements about the values of the $\overline{f}$ function. This establishes limitations on the formal axiomatic approach to mathematics. And, although the function $\overline{f}$ has a very abstract, non-constructive definition, it *is* a legitimate function (we don't know its values, but "The Universe" does). So, the inability of any formal system to derive its values shows that the formal axiomatic approach to discovering and verifying mathematical truths is inherently limited. This was the main philosophical impact of Gödel's First theorem. So although we have proven a weaker version of his theorem, we have come to the same philosophical conclusion.

Be careful not to misinterpret these results It is easy to (mis)state Theorems 6.5.1 and 6.5.2 (and alternate statements of them) as: There are true $\overline{f}$-statements that cannot be derived in any formal system. And that misstatement leads to statements such as: There are mathematical truths that cannot be proved in any formal system. Or even worse: There are mathematical truths that cannot be proved.

None of the three statements in the previous paragraph is correct. Consider the first one. It is not correct because it has *reversed* the direction of implication (for want of a better term). What is correct is that in any sound $\overline{f}$-expressible formal system, Π, there must be some true $\overline{f}$-statements (which ones depend on the particular Π) that cannot be derived in Π.

Review question Explain why the second and third statements above are also incorrect.

[23] Although, it also needs the minor qualifier that the system can derive some basic facts about arithmetic – a detail that we are leaving vague.

6.6.1 We Are Not There Yet

Theorems 6.5.1 and 6.5.2 carry (in my opinion) the most important philosophical force of Gödel's First incompleteness theorem (semantic version), but they are technically weaker. So, we want to keep the general outline of their proofs, but make changes that move the resulting theorem closer to what is actually in Gödel's First theorem (semantic version).

What do we want to improve? There are three ways we would like to strengthen the theorems.

First, the purely conceptual, abstract way that function $\overline{f}$ is defined makes the theorems less compelling than we want. Function $\overline{f}$ is defined only *in relation* to the set of all computable binary functions (which exists) and an ordering of them (which also exists). But we only know that the set of all computable functions exists, and that there exists an ordering of the computable functions. We don't know any way to actually identify functions that are computable, and we don't know any constructive way to order them.

Theorems 6.5.1 and 6.5.2 are correct, but their importance would be more convincing if instead of using computable functions and the function $\overline{f}$, we could use a set of objects whose definition is more *constructive*, fully independent of other objects, and where there is a constructive method to order the objects. We will make that improvement in Chapter 7, when we examine impossibility theorems of Alan Turing, specifically, the undecidability of the *Halting Problem.*

Second, although *soundness* is a natural and desirable property for a formal system, the *syntactic* version of Gödel's First theorem was eventually proved (by Rosser) with only the assumption of *consistency*[24] – a weaker assumption than soundness, so a stronger resulting theorem. A consistent system need not be sound, but a sound system will always be consistent. Proving a theorem starting from a weaker assumption means proving a stronger theorem, so the syntactic version of Gödel's First theorem is stronger than the semantic version.

We would like to prove, as Gödel/Rosser did, an incompleteness theorem that only assumes *consistency* rather than *soundness.* We will do this in the next chapter, Chapter 7, where we use Turing undecidability to prove incompleteness, assuming only consistency. We will also discuss the way that

[24] From a purely technical standpoint, that is really amazing and head–spinning (to me). It implies that even without thinking about "truth" or "meaning", just considerations of the syntax of a language and of the allowed rules of logic can lead to the provable incompleteness of a system.

soundness is replaced with *consistency* in Chaitin's incompleteness theorem, in Chapter 8, although that will not be a full proof.

Finally, in this chapter we only proved *existence* theorems. We proved that for any sound $\overline{f}$-expressible formal system Π, there *must exist* true $\overline{f}$-statements that are not derivable in Π. That is an *existence* proof. But Gödel did not just prove existence. He showed how to *construct*, for any "rich-enough" formal system Π, a statement called $G_T(\Pi)$ (which is dependent on the details of Π) where neither $G_T(\Pi)$ nor its negation can be derived in Π.

Statement $G_T(\Pi)$ is called a *Gödel sentence*, which Gödel used in the original proofs of both his First and Second incompleteness theorems. Gödel sentences have been *likened* to paradoxical English sentences that say "This statement is false". We will explicitly discuss Gödel sentences in Chapter 9.

6.6.2 And What Else?

In addition to explicit *self-reference* (mentioned earlier), and *Gödel sentences*, there is another key concept in Gödel's original proof that is missing from the discussion in this chapter. That is the concept of *Gödel numbering*, which will also be discussed in Chapter 9.

6.7 Exercises

1. In program P^*, what is the point of requiring program P to generate strings in order of their lengths, breaking ties lexicographically? Would the given proof of (our variant of) Gödel's theorem work if P did not generate the strings in that order, but could (somehow) generate all the strings in no predictable order?
2. Suppose program P generated strings in order of their length, but broke ties in no predictable order, rather than using lexicographic order to break ties. Would the proof of our variant of Gödel's theorem still work?
3. Show that Theorem 6.5.1 implies Theorem 6.5.3. What about the other direction?
4. Fully explain why a formal system that is sound must also be consistent.

7

Turing Undecidability and Incompleteness[1]

In this chapter, we first discuss and prove a theorem due to Alan Turing (the father of computer science and the most important architect of its theoretical foundations) which shows that for a specific problem, the *Halting problem*, it is *impossible* to write a computer program that is *guaranteed* to correctly solve *all* instances of the problem. Moreover, this is just one of a large number of real problems that have that property. Such problems are called *undecidable* or *unsolvable*.[2] After discussing undecidability, we will show its relationship to logic, Gödel and Gödel-ish *incompleteness*, both in terms of soundness and consistency.

7.1 Undecidable Decision Problems

Recall from Chapter 6 that a *function* f is called *computable* if there exists a computer program that can correctly compute the value, $f(x)$, for every input, x. A function that does not have this property is called *non-computable*. Now, we shift our focus from *functions* to *problems*, but the phenomena of interest are very related.[3]

[1] Compared to the other topics in this book, accessible yet rigorous mathematical expositions of Turing undecidability are much more widely known. This is because most computer science students (both undergraduate and graduate students) are required to study this material, and typical courses develop full mathematical proofs. However, unlike most expositions of Turing undecidability, ours is based on computer programs rather than *Turing machines*, simplifying the exposition for an audience wider than computer science students and professionals.

[2] There are several other equivalent terms that are commonly used, but I think two are enough – and some of the more mathematical terms are not even suggestive of their actual meanings.

[3] We could do the exposition in this chapter in terms of functions instead of problems, but it would be awkward and non standard.

What is a problem, and what is a problem instance? A *problem* is specified by a general (somewhat abstract) defining statement. In contrast to a general problem statement, an *instance* of a problem has all the needed problem details, so that a solution to the instance exists.[4] A *decision problem* is one where the solution (or answer) can *only* be *Yes* or *No*. For example, we can define the *Square Problem*, a decision problem, with the following defining statement:

Given an integer x, is x the square of some integer?

That general problem statement is not an instance of the Square Problem, since no value of x is specified. An *instance* of the *Square Problem* is created when a value for x is specified. The following is an instance of the Square Problem:

Is 4 the square of some integer number?

The solution (answer) to this problem instance is clearly *Yes*. But the solution to the following problem instance:

Is 7 the square of some integer?

is *No*.

Decidable problems If there is a computer program that can correctly solve *every* instance of a particular decision problem, that problem is called *decidable* or *solvable*. Otherwise, that problem is called *undecidable* or *unsolvable* (duh!).

Review question: The *Square Problem* is clearly a decidable problem. Explain why.

7.2 The Halting Problem Is Undecidable

In Chapter 6, we proved that there are *non-computable functions*. In this section, we will prove that there are *undecidable problems*, but the proofs follow almost identical logic. We will first examine the problem called *The Halting Problem*, which was central in Alan Turing's development of the theory of *undecidable problems*.

[4] Although, it may be quite difficult to find the solution.

Recall from Chapter 6 that we can recast any computer program, p, as a *single* long *string*, s, simply by concatenating (joining) the statements in program p (assuming each statement has an end-of-line marker). String s is called the *string-encoding* of program p. We can now introduce

The Halting Problem: Given a program p (specified by the string-encoding, s, of p), and given a string w (which might be the *empty-string*, that is, no string), will the execution of Program p ever *halt* (terminate) if p is given string w as input?

An *instance* of the Halting Problem, denoted (s, w), is specified when values for strings s and w are given. The Halting Problem is a decision problem since the answer for any given instance is either *Yes* or *No*.

The Halting Problem is decidable if and only if there is some computer program, call it *PH*, that can correctly solve every instance (s, w) of the problem.

7.2.1 Not Halting Is (Usually) Not Good

An execution of a program that never halts will run forever, which is generally not what the designers of the program want. So, a program that does not halt on some input indicates that the program has a significant bug (error). And, such bugs are common. Hence, the *Halting Problem* is a real problem, and a program that could correctly solve every instance of it would have great practical value to all programmers, and also great monetary value to many software companies.

And more profoundly If there were a program *PH* that could correctly solve every instance of the Halting Problem, that would have profound impact on many open (unanswered) questions in mathematics. To see why, consider the famous open question called *Goldbach's Conjecture*.

Goldbach's conjecture is that *every even* positive integer larger than two is the sum of two *prime* numbers (possibly the same prime).[5] Three examples: $12 = 7 + 5$, $4 = 2 + 2$, and $102 = 97 + 5$, where 4, 12, and 102 are even positive numbers, and 2, 5, 7, and 97 are prime numbers.

Goldbach's conjecture was stated in 1742 and the great mathematician Leonhard Euler wrote: "That every even integer is a sum of two primes, I regard as a completely certain theorem, although I cannot prove it." It has been verified by computer for all even integers up to the cosmic number of 4×10^{18}. Despite all efforts over the last 280 years, it remains *unproven* and *unrefuted*. No one knows for sure whether it is true or false. But, if there were a

[5] Remember that a *prime* number, n, is a positive integer where the only positive integers that evenly divide n (i.e., with no remainder) are 1 and n.

program, *PH*, that could correctly solve every instance of the Halting Problem, we could use *PH* to *resolve* Goldbach's conjecture.

Here is the idea As noted above, there is a computer program, call it *PG*, that successively generates the even positive integers, and for each generated integer, it determines whether or not that integer is the sum of two prime numbers (such a program is conceptually easy to design). We design program *PG* so that if it ever examines an even integer, n, and determines that n is *not* the sum of two prime numbers, it will halt, having established that Goldbach's conjecture is *false*. Otherwise (in the case that Goldbach's conjecture is actually true), Program *PG* will *never* halt – it will continue to generate and examine the even positive integers. So, the question of whether Program *PG* ever halts is equivalent to the question of whether Goldbach's conjecture is true or false.

Given the observation in the previous paragraph, if there were a program *PH* that could correctly solve every instance of the Halting Problem, we could resolve Goldbach's conjecture by giving Program *PH* the input (s, w), where s is the string-encoding of Program *PG* and w is the *empty string*. With that input, if Program *PH* correctly determines that Program *PG* will halt, then Goldbach's conjecture must be *false*; and if Program *PH* determines that Program *PG* will *not* halt, then Goldbach's conjecture must be *true*.

So, the existence of a Program *PH* would resolve Goldbach's conjecture, having a very significant impact in mathematics. And, there are many other open problems in mathematics that could be resolved in this manner. This makes it easier to *believe* that the existence of such a Program *PH* is *impossible*. But a belief is not a proof. What we want is a rigorous proof that Program *PH* is impossible. We will shortly develop such a proof, due to Alan Turing.

But wait! Why isn't it trivial? At first, it may seem that it is trivial to correctly solve any instance of the Halting Problem: Given any specific instance (s, w), where s is the string-encoding of a Program p and w is the input to Program p, why not just let Program p run with input w? If that execution of p halts, then Program *PH* knows that the solution to this instance of the Halting Problem is *Yes*, and if p doesn't ever halt, then .... (??). Oh, there is the rub!

If Program p, given input w, never halts, how will Program *PH* know that? Program p may have run for a long time, but how long should Program *PH* wait before declaring that p will *never* halt? *We* don't know how long to wait, so what time limit can we build into Program *PH*? If we wait 6,000 years and then declare that the execution of p will never halt, we might be wrong – it might halt if we just allowed one more second of execution.

We don't know a way around this issue – although not knowing a way around it is *not* a proof that there is no way. But in fact, there really is *no* way around this "how-long-to-wait" issue. That is because we will *prove* that the Halting Problem is *undecidable*, that is, that it is *impossible* to create a computer program that correctly solves *all* instances of the Halting Problem. For that proof, we will use a *restricted version* of the Halting Problem.

7.2.2 The Self-Halting Problem[6]

When trying to prove that a problem is *undecidable* (rather than decidable), it is logically acceptable, and often helpful, to restrict attention to a *subset* of all possible instances of the problem. If we prove that the problem is undecidable when it is *restricted* to a subset of instances, that also implies that the general problem is undecidable.

In particular, we will restrict the Halting Problem to instances (s, w) where the string-encoding, s, and the string, w, are *identical*. That is, we specialize the Halting Problem to instances of the form (s_i, s_i), where s_i is the string-encoding of a program P_i. This is called the *Self-Halting* problem. Since the two strings are identical, an instance of the Self-Halting problem can be specified by a single string, s_i.

Recapping, the Self-Halting Problem is the decision problem:

> Will Program P_i ever halt if run with input consisting of its *own* string-encoding, s_i?

7.2.2.1 The Analysis of the Self-Halting Problem

We begin our analysis of the Self-Halting Problem by observing (similar to what we did in Chapter 6) that it is possible to *order* all the computer programs (in some chosen programming language $\mathcal{L}_\mathcal{P}$), into a list, $L_\mathcal{P}$. As in Chapter 6, we see this by imagining first an ordered list, L_S, of all the *strings* created in the alphabet of that programming language. The strings in L_S are ordered by their lengths (shortest first), breaking ties lexicographically for strings of the same length. Then, a *subset* of the strings in L_S are string-encodings of computer programs (in the chosen language), showing that there exists an ordering of all computer programs into the list $L_\mathcal{P}$.

Table *HP* showing Self-Halting behaviors With list $L_\mathcal{P}$, we can conceive of a table, call it Table *HP*, showing the Halting/non-Halting behaviors of each

[6] I often misread this as the *Self-Hating* Problem. Do you? Does this have any psychological meaning?

Table 7.1 *The conceptual Table HP has an ordered listing of all computer programs in a specified programming language, on the vertical axis, and the string-encodings of all the computer programs in the same order, on the horizontal axis. But only the main diagonal is of interest. Each cell, (i, i), on the main diagonal has a value, either H or nH, showing whether or not Program P_i halts when it is executed with input s_i, that is, its own string-encoding.*

HP	s_1	s_2	s_3	s_4	s_5	. . .	s_i	...
P_1	H							
P_2		nH						
P_3			nH					
P_4				H				
P_5					nH			
.						.		
.						.		
.						.		
P_i							H or nH	
⋮								⋱

program in $L_{\mathcal{P}}$ when given their own string-encoding as input. An example is shown in Table 7.1. Table *HP* is similar to Table *T* discussed in Chapter 6, but now we are only interested in the cells on the main diagonal of the table.

Self-Halting is undecidable We want to prove that the Self-Halting Problem is *undecidable*, which we will do with a *proof by contradiction*.

Suppose the Self-Halting Problem *is* decidable, so there *is* a program, let's call it *SHP*, that can correctly answer any instance of the Self-Halting Problem. Program *SHP* takes as input a single string-encoding, s_i, for some program P_i, and correctly answers the question:

> Will program P_i (which can be obtained from s_i) halt when executed with input consisting of string s_i?

Note that in order to correctly solve instances of the Self-Halting Problem, the hypothetical Program *SHP* must itself terminate (halt), given any input and it must always produce an output, either *Yes* or *No*. Table 7.2 shows the *outputs* of (the hypothetical) Program *SHP*, along the main diagonal. Table 7.2 is similar to Table 7.1, but every entry *"H"* has been replaced by *"Yes"*, and every entry *"nH"* has been replaced by *"No"*. Although this is a trivial modification, using a different table makes explicit that the entries in Table 7.2

Table 7.2 *The conceptual Table SHP has an enumeration, $L_{\mathcal{P}}$, of all computer programs in a specified programming language (on the vertical axis), and the string-encodings of all the computer programs, in the same order, on the horizontal axis, but only the main diagonal is of interest. Each cell, (i, i), on the main diagonal has a value, either Yes or No, showing what the (hypothetical) Program SHP would output when given the string-encoding s_i for program P_i.*

SHP	s_1	s_2	s_3	s_4	s_5	. . .	s_i	...
P_1	Yes							
P_2		No						
P_3			No					
P_4				Yes				
P_5					No			
.						.		
.						.		
.						.		
P_i							Yes or No	
⋮								⋱

are the outputs of the *single* Program *SHP*, while the entries in Table 7.1 are the *behaviors* of each programs P_i when given input s_i. To make the relationship between Tables 7.1 and 7.2 more explicit, we restate it as:

Theorem 7.2.1 *The* (i, i) *entry in Table 7.2 has value Yes if and only if the* (i, i) *entry in Table 7.1 has value H, if and only if Program* P_i *halts when executed with input consisting of its own string-encoding,* s_i.

Program SHP-NOT With the assumption that Program *SHP* exists, we will *design* another program, call it *SHP-NOT*, based on Program *SHP*. Program *SHP-NOT* takes in the same input as Program *SHP*, that is, the string-encoding, s_i, of some program P_i.

We design Program *SHP-NOT* so that when it is given input s_i, it will first run Program *SHP* with input s_i. By assumption, that execution of Program *SHP* will determine whether (or not) program P_i will halt when given input s_i, outputting either *Yes* or *No*. But, we design Program *SHP-NOT* to *intercept* the output of Program *SHP* before the output reaches the outside world.

When Program *SHP-NOT* intercepts the output of Program *SHP*, Program *SHP-NOT* does the following (See Figure 7.1):

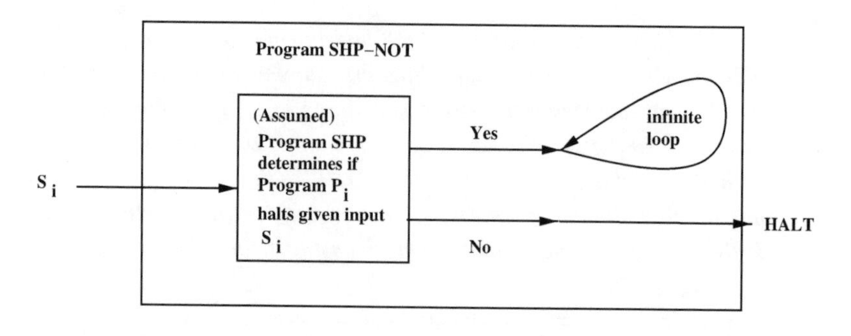

Figure 7.1 Program *SHP-NOT*, built from the Program *SHP*, which we assumed to exist.

If the output from Program *SHP* is *Yes*, then Program *SHP-NOT* will enter into an infinite loop, and *never halt*.

But if the output from Program *SHP* is *No*, then Program *SHP-NOT* will *immediately halt*.

Note that Program *SHP-NOT* does not produce any output – its only signal to the outside world is its halting, if it halts.

Now that you see the details of Program *SHP-NOT*, it should be clear that *if* Program *SHP* exists, *then* Program *SHP-NOT* also exists, since it is an easy modification of Program *SHP*.

Recapping: Assume Program *SHP* exists. Then Program *SHP-NOT* also exists. Program *SHP-NOT*, when given input s_i, will halt if and only if Program *SHP*, when given s_i, will output *No*; if and only if cell (i, i) of Table *HP* (as in Table 7.1) has value *nH*; and if and only if Program P_i will *not* halt, when given input s_i. Removing the middle statements, we get

Theorem 7.2.2 *Assuming Program SHP exists, Program SHP-NOT also exists, and will halt when given input s_i, if and only if Program P_i will not halt, when given s_i.*

OK, this is a bit much to absorb quickly. It helps to alternate between reading the statement of this theorem and looking at Figure 7.1.

Review question: Is the following argument correct? Explain.
Assume Program *SHP* exists. Program *SHP-NOT*, when given input s_i, will not halt if and only if Program *SHP*, when given s_i, will output *Yes*; if and only if cell (i, i) of Table 7.1 has value *H*; and if and only and if Program P_i will halt, when given input s_i.

7.2.2.2 And Now for the Kicker

Theorem 7.2.3 *It is impossible to create a program, SHP, that can correctly solve all instances of the Self-Halting Problem. That is, the Self-Halting Problem is undecidable.*

Proof Assume, for a proof by contradiction, that Program *SHP* does exist. Then Program *SHP-NOT* also exists, and so it *must* be somewhere in the list of programs $L_{\mathcal{P}}$, and it must be represented by some row of Table 7.1. But, which row?

As a warm-up, let's (arbitrarily) see if it could be row 57. If it is, then Program *SHP-NOT* must be Program P_{57} in the list $L_{\mathcal{P}}$. But, by Theorem 7.2.2 Program *SHP-NOT will* halt, when given string s_{57} as input, if and only if Program P_{57} will *not* halt, when given string s_{57} as input. Hence, programs *SHP-NOT* and P_{57} have *different* behaviors when given string s_{57} as input, and so they *cannot* be the same program. So, Program *SHP-NOT* is not represented by row 57 of Table 7.1.

To fully prove the theorem, replace P_{57} with P_i in the above argument. The argument still works, so Program *SHP-NOT* cannot be Program P_i, for any i, and hence cannot be in the list $L_{\mathcal{P}}$. Therefore, Program *SHP-NOT* cannot exist.

So, the assumption that Program *SHP* exists leads to the conclusion that Program *SHP-NOT* does *not* exist. The only resolution to this contradiction is that the initial assumption (that Program *SHP* exists) is wrong. So Program *SHP* does *not* exist, meaning that the Self-Halting Problem is undecidable. ■

And, as we argued earlier, because the Self-Halting Problem is a restriction of the general Halting Problem, we can also conclude that the general *Halting Problem* is undecidable.

7.2.3 Turing's Proof in Rhyme

Many people have admired the cleverness and beauty of Turing's proof and have come up with their own expositions to make it more accessible to the general public. The most notable of these is a Dr. Seuss-style poem created by Dr. Geoffrey K. Pullum, which works its way through Turing's proof of the undecidability of the Halting Problem in 14 stanzas. Below are the first two stanzas, and the last two stanzas of the poem.[7]

No general procedure for bug checks will do.
Now, I won't just assert that, I'll prove it to you.

[7] Copyright ©2008, 2012, 2022 by Geoffrey K. Pullum, used by permission of the author.

I will prove that although you might work till you drop,
you cannot tell if computation will stop.

For imagine we have a procedure called P
that for specified input permits you to see
whether specified source code, with all of its faults,
defines a routine that eventually halts.

...

So where can this argument possibly go?
I don't have to tell you; I'm sure you must know.
We've proved that there just cannot possibly be
a procedure that acts like the mythical P.

You can never find general mechanical means
for predicting the acts of computing machines;
it's something that cannot be done. So we users
must find our own bugs. Our computers are losers!

For the complete proof (poem), see the webpage cited in [87] or page 196 in [62].

7.2.4 Is the Halting Problem Special?

We have proved that the Halting problem is undecidable, and we earlier proved the related result that some functions are non-computable. That's interesting, but maybe these are just odd, isolated problems and functions.

No, they aren't!

A HUGE number of known functions are non-computable, and a HUGE number of interesting, important problems are undecidable. Another undecidable problem – The *program equivalence* problem – is discussed in the next section. A wide variety of undecidable problems and non-computable functions of interest in computer science are discussed in many textbooks, such as [97].

There are also many undecidable problems outside of computer science. One of the most visually compelling problems concerns Conway's famous *Game-of-Life* (see [107] for an introduction and several moving illustrations). In this undecidable problem, one is given two configurations, and the decision question is whether there is an execution of the game-of-life that starts with the first configuration and leads eventually to the second configuration. The question is undecidable – there is no algorithm that can decide that question for all pairs of configurations.

Other significant undecidable problems outside of computer science include the undecidability of the spectral-gap problem [28, 29, 66, 56] and the non-computability of questions about the wave function [86], in quantum mechanics; additional undecidable problems in quantum mechanics [34, 114, 84]; the undecidability of questions about dynamical systems [75] in general physics; the undecidability of questions about infinite tilings [90] in mathematics; and most notably, the undecidability of the question of whether a given polynomial equation has a solution only using integer values [30].[8]

A list of important undecidable problems in several branches of mathematics appears in [85]. That paper also cites other papers that give lists and reviews of undecidable problems. In addition to its main result mentioned above, many undecidable problems in physics are discussed in [28] (see also, someday hopefully soon, [27]).

Additionally, there are general theorems that characterize *infinite-sized* sets of problems that are undecidable. The best-known such theorem is called *Rice's* theorem. Its proof is not more difficult than what we have already proven, and we leave it to the interested reader to look it up.

7.2.4.1 Another Important Undecidable Problem

We briefly discuss another extremely practical undecidable problem:

The Program-Equivalence Problem

> Given the string-encodings, s_i and s_j, of two programs, P_i and P_j, will the output of P_i be *identical* to the output of P_j, for *every* input, w, given to both P_i and P_j?

A program that could correctly solve every instance of the Program-Equivalence Problem would be extremely valuable for software developers. It often happens that one has an existing program P_i and wants to create an improved program (smaller, faster, or in another language) P_j whose input/output behavior is *identical* to program P_i. Or, a user has the choice of using two different programs that claim to compute the same things, and the user wants to be sure that they really do. Such situations occur frequently.

But, how can one be sure that two different programs are actually equivalent? If we had a master program that could correctly solves all instances of the Program Equivalence Problem, we could input two string-encodings, s_i and s_j, of two programs, P_i and P_j, into that master program, and learn whether or

[8] Such an equation is called *Diophantine*. The existence or nonexistence of an algorithm to solve all Diophantine equations is called Hilbert's 10th problem, because it was the 10th problem in the list of 23 open mathematical problems enumerated by David Hilbert in a famous lecture in 1900. The nonexistence of such an algorithm was finally proved more than 70 years later.

not the two programs were actually equivalent. Even if that master program was slow, because it would only need to be used infrequently, it would be extremely valuable.

Unfortunately, there is no such master program – it has been proven impossible. The Program-Equivalence Problem is undecidable. We will not prove that in this book, but its proof is no more difficult than our proof that the Halting Problem is undecidable.

7.3 Using the Halting Problem to Establish Incompleteness

In Chapter 1 (Introduction), we quoted comments of Gregory Chaitin [23] on Gödel's First incompleteness theorem, but we did not include the sentence that comes immediately after that quote. Here is a more complete quote:

> A great many different proofs of Gödel's theorem are now known, and the result is now considered easy to prove and almost obvious: It is *equivalent* to the unsolvability of the halting problem. (italics added)

In Chapter 1, I said that I disagreed that Gödel's theorem was *easy* to prove or almost *obvious*. Now I disagree with what is implied by the added sentence that Gödel's theorem is easy and almost obvious *because* it is equivalent to the unsolvability of the Halting Problem. In my opinion, the equivalence is neither easy nor obvious to prove.

Still, it is clear that Gödel's incompleteness theorem and Turing's undecidability theorem are *extremely* related, and undecidability leads easily to a *variant* of Gödel's incompleteness theorem. We show this next.

7.3.1 A Dramatic Turn

Our exposition follows the presentation in [81], which contains the following introduction:

> We now come to the dramatic turn in our story. Namely, we're about to demand that our formal system can reason about algorithms ...

We will interpret the word "algorithms" in the above quote as "programs", which is a more appropriate word since a program is a detailed description of an algorithm in a specified programming language, and a formal system needs such a full specification in order to derive statements about it.

Reasoning about programs But what does it mean for a formal system Π to "reason about programs"? The ability to reason about programs has several implications, but for now, we only need to discuss one.

Different programs have different well-specified *properties*. For example, some programs have the property of including a *FOR* loop and some do not. Some programs have the property of being more than 500 statements long, and some do not. And some programs have the property of halting when given their string-encoding as input and some do not. These are all well-defined properties of programs, although it is easy to determine if a program has either of the first two properties, while that is not true for the third property, since the Self-Halting Problem is undecidable.

With the notion of program *properties* defined, when we say that a formal system Π can *reason about programs*, we require that for any program, p, written in some computer language, and for any property, Z, the formal system Π should be able to *form* the statements that mean "Program p has property Z" and "Program p does not have property Z". Further, if formal system Π is *complete*, then it must be able to *derive* at least one of those two statements; and if Π is *consistent* then it must *not* be able to derive both of those statements; and if Π is *sound*, then it must *only* be able to derive whichever of the two statements is *true* (assuming that truth has been defined for Π).

Back to incompleteness Now consider the property: *Halts when given the string-encoding of itself*, which is a property that some programs have and some do not have. So, for any program P_i, a formal system that can reason about programs must be able to form the two statements, *RA* and *RB* that mean:

RA) Program P_i (specified by the input s_i) *halts* when given s_i as input.

And

RB) Program P_i (specified by s_i) does *not* halt when given s_i as input.

OK, so the assumed formal system can *form* statements *RA* and *RB*, but what can the system *derive* from its axioms, using its allowed inference rules?

Theorem 7.3.1 *No sound formal system* Π *that can "reason about programs" can be complete. That is, the system* Π *can form a statement such that neither it nor its negation can be derived in* Π.

Proof We will show that *if* Theorem 7.3.1 is *not* true, then the Self-Halting Problem would be decidable. But, since we have already established that the Self-Halting Problem is **un**decidable, Theorem 7.3.1 *must* be *true*.

Suppose that Theorem 7.3.1 is *not* true. That is, there is a sound, complete formal system Π that can reason about programs. So, for *any* statement that can be formed in Π, either it or its negation (but not both) can be derived in Π. Then, we will use this (supposed) system Π to create a computer program, *SHP-ALT*, that can correctly solve any instance of the Self-Halting Problem. Recall that an instance of the Self-Halting Problem is specified by a string s_i, where s_i is the string-encoding of a computer program P_i.

Program SHP-ALT Given the input, s_i, Program *SHP-ALT* will start generating all strings that can be formed using the alphabet of Π, in order of length, ordering equal-length strings lexicographically. (Such a generation of strings has been used several times in earlier chapters.) Each generated string is a potential derivation of some statement in Π. As discussed before (see Section 6.3.2), for any formal system Π, there is a computer program, *PC*, that can check a string to determine if the string is a formal derivation in Π. Program *SHP-ALT* will incorporate Program *PC* to check each generated string to see if it is a derivation of a statement, and if so, to check if the derived statement is one of the statements *RA* or *RB* listed above.

Further, given input s_i, Program *SHP-ALT* will alternate between generating a string and running Program *PC* to check whether the generated string is a derivation of statement *RA* or *RB*. Since system Π is assumed to be *complete*, a derivation of one of those two statements will eventually be generated and checked. And, since Π is assumed to be *sound*, the actual generated statement will be *true*, correctly stating whether or not Program P_i will halt when given s_i as input.

Summarizing: The original assumption, that Theorem 7.3.1 is *not* true, leads to the creation of Program *SHP-ALT* that can correctly solve any instance of the Self-Halting Problem. But, this is impossible, since we know that the Self-Halting Problem is *undecidable*. The only way to resolve this contradiction is to conclude that the original assumption in the proof must be wrong, so that Theorem 7.3.1 must be *true*. ■

An alternative proof of Theorem 7.3.1 will be given on page 166.

A Gödel variant What Theorem 7.3.1 establishes is that the *undecidability* of the Self-Halting Problem implies the *incompleteness* of any sound formal system that can "reason about programs" (or just reason about the Halting Problem, in particular). This is a *variant* of Gödel's First incompleteness theorem (semantic version). It differs from Gödel's actual theorem in that it assumes that the formal system has the ability to reason about *programs*, while Gödel's theorem only makes the (seemingly) weaker assumption that the system can reason about *arithmetic*. And, a weaker assumption means a stronger resulting theorem – which is part of why the proof of Gödel's actual first incompleteness theorem is more difficult than what we did in this chapter (and in Chapter 6). As stated in [81]:

> By placing computation ... directly into the bedrock of our formal system our task [becomes] orders of magnitude simpler.

Actually, a Gödel equivalent Although we will not show this,[9] the feat of reasoning about *programs* can be accomplished by systems that only reason about *arithmetic*. Once that is established, Theorem 7.3.1 *really does* imply Gödel's first incompleteness theorem (semantic version), and not just a Gödel-*ish* theorem.

7.3.2 Isn't This Dèjá vu All Over Again?

The proof given for Theorem 7.3.1 follows that same general outline as the proof of Theorem 6.5.1 in Chapter 6. The major difference is that the proof of Theorem 6.5.1 used the *non-computability* of *function* $\overline{f}$, while the above proof of incompleteness uses the *undecidability* of the Self-Halting *Problem*. Why is that better?

To answer this question, recall from Section 6.6 that there are several ways we wanted to improve Theorem 6.5.1 (and its friends). The proof based on the Self-Halting Problem addresses the first, and most important, of these.

Construction vs. existence Theorem 6.5.1 is based on an ordering of the *computable functions*. But the list, L_C, of computable functions was only known to *exist* – we didn't have any way to *construct* that list – so it was very abstract. As one reflection, if we are given a function (somehow defined) and want to determine if it is computable, we have no tools to address that question. We have no way to guarantee a *Yes* answer when the correct answer is *Yes*, or to guarantee a *No* answer when the correct answer is *No*.

In contrast, if we are given a program, P_i, and want to know if it halts when given its string-encoding, s_i, as input, we can just execute P_i with input s_i. If in fact P_i halts when given its string-encoding, then that execution will eventually halt, and we will know that *Yes* is the correct answer to that instance of the Self-Halting problem.

[9] Because that is where grunge lives.

Of course, if P_i *never* halts, we will *not* know the answer to the question, and we do *not* have a method to guarantee a *No* answer when *No* is the correct answer. Still, what we can constructively determine about Self-Halting is much more than we can constructively determine about computable functions as a whole. For a discussion on the conceptual importance of constructions in mathematics, versus pure existence proofs, see [21].

More construction Further, we can *constructively* create the list of all programs, $L_{\mathcal{P}}$, in the order that the programs are in the list of all strings. The *strings* using the alphabet of the chosen programming language can be generated by a computer program, in order of length (and lexicographically among strings of the same length), although the program will run forever because there are an infinite number of strings. Also, the generated strings that are string-encodings of computer programs can be identified by a computer program called a *compiler*, which is routinely used to determine if a given string s is the string-encoding of a grammatically correct, legal program in the chosen language. Hence a string generating program, alternating with a compiler that checks for a legal program, can *constructively* generate, in order, the list $L_{\mathcal{P}}$.

Even more construction Even further, we can *constructively* create a list, call it L_H, of all programs (in the chosen programming language) that halt when given their own string-encodings as input.

Let's try an approach that will actually fail: After each program P_i is identified and added to list $L_{\mathcal{P}}$, as described above, execute Program P_i with its own string-encoding, s_i, as input. As discussed earlier, if P_i halts when given s_i as input, this execution will halt and then P_i can be added to the list L_H. That sounds good, but we said it will fail. What's the problem?

Review question Before reading further, answer the question: What is the problem?

The problem occurs the first time the process generates and starts executing a program P_i that *never* halts when given its own string-encoding as input. At that point, the above method will get stuck executing P_i forever, and no more programs will be added to list L_H. But, there is actually a clever fix to this problem.

A fix When we add a new program P_i to $L_{\mathcal{P}}$, we start an execution of P_i with input s_i, but we *only* let that execution take *one* step. After that one step, we pause the execution of P_i and start a *round* through the programs on list $L_{\mathcal{P}}$ whose executions have *not* terminated. In a single round, we let each such program take one more step in its execution. Any program that terminates in that round is inserted into list L_H.[10]

When a round is finished, we generate the next program for list $L_{\mathcal{P}}$, let it execute its first step (with input consisting of its own string-encoding), pause it, and start a new round through all the programs on $L_{\mathcal{P}}$ that have not terminated. In this way, any program that halts when given its own string-encoding as input will eventually halt in this process and be added to L_H after a *finite* number of rounds. Of course, this process never ends, but it always makes progress – it won't get stuck forever on a program that doesn't terminate. This clever process is called *dovetailing*.

7.4 Advanced Topic: From Soundness to Consistency[11]

Theorem 7.3.1 established: Any *sound* formal system Π that can "reason about programs" must be incomplete. But soundness is a semantic property involving the concept of "truth". Gödel wanted to avoid any reliance on notions of truth and only use the *syntactic* property of *consistency*, that is, the avoidance of any *contradictory* statements. As we discussed in Chapter 6, consistency is a pretty weak requirement compared to soundness, so when a theorem is only based on the assumption of consistency, rather than on the assumption of soundness, it is a much stronger (and impressive) result. So, we would like to improve Theorem 7.3.1 to:

Theorem 7.4.1 *Any* consistent *formal system* Π *that can "reason about programs" must be incomplete.*

Before proving this theorem, we will present a second proof of Theorem 7.3.1. Then, an additional twist to that proof will lead to a proof of Theorem 7.4.1. We first describe Program *ISHP-ALT*, which is a modification of Program *SHP-ALT*.

Program ISHP-ALT As in Program *SHP-ALT*, when given the input s_i, Program *ISHP-ALT* will start generating all strings that can be formed using the

[10] We say that programs are *inserted* into L_H, and not *added* to L_H, because the order that program executions terminate may be different from the ordering of programs in L_H by their lengths, and lexicographically for programs of the same length.

[11] The material in this section is more advanced, and somewhat more difficult than the preceding material in this chapter. But, it requires no more mathematical knowledge than does the preceding material – it requires only careful reasoning. The exposition is based on the proof of Theorem 5 in [81].

alphabet of Π, in order of length, ordering equal-length strings lexicographically. And, as in Program *SHP-ALT*, each generated string will be checked by Program *PC* to see whether the generated string is a derivation of one of the statements *RA* or *RB* (given on page 161). Program *ISHP-ALT* differs from Program *SHP-ALT* in what happens *if* it does generate and identify a derivation of statement *RA* or *RB*.

If Program *ISHP-ALT* finds a derivation of statement *RB*, it will *immediately halt*. But, if Program *ISHP-ALT* finds a derivation of statement *RA*, it will go into an *infinite loop*, never halting. In more detail (and we will see later why it is important to specify such detail), we will assume that to enter into an infinite loop, Program *ISHP-ALT* executes the following instructions:

Infinite loop

repeat { set variable *looping* to *true* }
until { variable *looping* is *false* }

Clearly, once the program starts executing this loop, it will never halt, since the variable *looping* always has value *true*, but the loop will only terminate if the value becomes *false*. Note that if Π is *complete*, Program *ISHP-ALT* will either find a derivation of statement *RA* or statement *RB*. So, the only way that it fails to halt is by going into the Infinite loop described above.

Modified statements *RA* **and** *RB* We will use the notation *SI* to refer to the string-encoding of Program *ISHP-ALT*. Focus on an instance of the Self-Halting Problem where the input is *SI*. For this instance, Statements *RA* and *RB* now specialize to:

IA) Program *ISHP-ALT* (specified by the input *SI*) *halts* when given *SI* as input.

And

IB) Program *ISHP-ALT* (specified by *SI*) *does not halt* when given *SI*.

Note that Statement *IB* is the negation of Statement *IA*, and similarly, *IA* is the negation of *IB*. Formal system Π can form both statements *IA* and *IB* since it can "reason about programs". But, by the details of Program *ISHP-ALT*, if Π can *derive* statement *IA* or *IB*, it cannot derive the other statement. We are now ready for the

Second Proof of Theorem 7.3.1: We will assume that Π is *complete* and show that this leads to the conclusion that Π is *not sound*, so Π cannot be both complete and sound. Since Π is complete, a derivation of at least one of the statements *IA* or *IB* will be generated and found by Program *ISHP-ALT*. So, the only way that Program *ISHP-ALT* can fail to halt is if it goes into the Infinite loop described earlier.

There are two cases to consider:

Case 1: A derivation in Π of Statement *IB* is found during the execution of Program *ISHP-ALT* with input *SI*. Statement *IB* says that Program *ISHP-ALT* does *not* halt when given input *SI*. But, by the details of *ISHP-ALT*, when the program generates a derivation of Statement *IB*, it then *does* halt. And so, in this case, the formal system Π is *not* sound, since it derives a statement that is *false*.

Case 2: A derivation in Π of Statement *IA* is found during the execution of Program *ISHP-ALT* with input *SI*. Statement *IA* says that Program *ISHP-ALT does* halt when given input *SI*. But, by the details of *ISHP-ALT*, when the program generates a derivation of Statement *IA*, it then goes into an *infinite loop*, never halting. And so, in this case, the formal system Π is *not* sound, since it derives a statement that is *false*.

So, in both cases, the assumptions that Π can reason about programs, and that it is complete, leads to the conclusion that Π is not sound, giving a second proof of Theorem 7.3.1. ■

An interesting point about this proof is that it does not rely on the undecidability of any problem. It is completely self-contained.

7.4.1 On to Consistency

We will prove Theorem 7.4.1 by modifying the above proof of Theorem 7.3.1. This modification requires specifying more detail about what it means for a formal system to "reason about programs".

More on reasoning about programs First, the programs that Π reasons about are written in some specific programming language, abstractly denoted $\mathcal{L}_{\mathcal{P}}$. So, for Π to be able to reason about programs written in $\mathcal{P}$, the language of Π must *contain* $\mathcal{L}_{\mathcal{P}}$. Therefore, any specific program p written in language $\mathcal{L}_{\mathcal{P}}$ can be written in the language of Π.

Second, for any program p written in $\mathcal{L}_{\mathcal{P}}$, and any input I, the formal system Π must be able to "step through" or "describe" or "explain" any *finite* number of steps of the execution of program p with input I.[12]

[12] This is a subtle point that is typically left a bit vague. For example, [81] explains the relationship of a computation of a program (written in language $\mathcal{L}_{\mathcal{P}}$) to a derivation (written

The idea is that each step in the execution of program p (given input I) is determined by a specific statement in program p, along with the values of the variables in p when that statement is executed. For example, the statement in p might compare the value of some variable to a constant number, and the outcome of that comparison then determines what the next step in the execution of p will be. This means that the action of each step of the execution of program p, given I, can be "described" or "explained" by statements in Π that parallel the execution of the program. In this way, each step in the execution of p (with an input I) can be turned into several statements in Π explaining that step. One way to think about this is that statements in Π provide a step-by-step "running commentary" on the execution of program p with input I.[13]

Now here is the **key point:** The running commentary that Π provides on a specific execution of program p, given input I, is also a *derivation* in Π of what program p computes, given input I. When we say that Π can "reason about programs", it implies that Π can provide such a running commentary, and this commentary provides derivations of statements that mean "The execution of Program p with input I, does the following: ... ". So, if an execution of Program *ISHP-ALT*, with input *SI*, ends by halting, the running commentary provides a derivation in Π of Statement *IA*. Similarly, if an execution of Program *ISHP-ALT* enters the infinite loop detailed above, then the running commentary provides a derivation in Π of Statement *IB*. With that insight, we can now proceed to a proof of Theorem 7.4.1 (stated on page 165).

Proof of Theorem 7.4.1: Remember that the assumption that Π is consistent is part of the premise of the theorem. In addition, for a proof by contradition, we will assume that Π is also complete, and then derive a contradiction. The contradiction means that Π cannot be both consistent and complete.

Because Π is assumed to be complete, and statements *IA* and *IB* are negations of each other, any execution of Program *ISHP-ALT* will generate and identify a derivation in Π of exactly one of those two statements.

In Case 1, the execution ends by *generating* and *identifying* a derivation in Π of statement *IB*. But then, the running commentary that Π provides for this

in the language of Π) that describes that computation, simply as: "We can write out this computation inside Π." I think such a vague statement might be sufficient for people who have good familiarity with computer programming, and the intuition that one develops from programming. But, here I will try to make this relationship more explicit. Also, if you do not have any familiarity with computer programming languages, please see the Appendix for a light introduction.

[13] If you have done some computer programming, you probably have "stepped through" an execution of a program, with some given input, to explain the workings of the program to someone else, or while debugging the program.

execution is actually a derivation in Π of the assertion that Program *ISHP-ALT* *halts*, when given input *SI*. Similarly, a derivation of the assertion that Program *ISHP-ALT* halts, when given input *SI*, is a derivation of Statement *IA*. So, in Case 1, there will be a derivation in Π of both statement *IB* and Statement *IA*.

This argument is a bit subtle, so I will add some more detail. In Case 1, an *explicit* derivation in Π of Statement *IB* is found by the execution of Program *ISHP-ALT*, generating all strings that can be written in Π, checking each one to see if the string is a derivation of Statement *IA* or *IB*, and finding that a generated string is a derivation of Statement *IB*.[14]

But in Case 1, in addition to what is found by the execution, the *running commentary* for that execution provides a *derivation* in Π of Statement *IA*. So, in Case 1, we see that Π can derive a statement (*IB*) and also derive its negation (*IA*). Summarizing, we have proved that in Case 1, there are derivations in Π of both Statements *IA* and *IB*, and so in this case, the formal system Π is *inconsistent* contradicting the premise of theorem.

Note that the derivation of Statement *IA*, obtained from the running commentary, must be longer than the derivation of Statement *IB*, since the *IB* statement was assumed to be found first in the execution of Program *ISHP-ALT*.

The analysis of Case 2 is similar to Case 1, but there is an added subtlety. In Case 2, the execution of Program *ISHP-ALT* never terminates but is caught in the Infinite loop described earlier. The Infinite loop is entered after the generation and identification of a derivation in Π of statement *IA*.

I claim that the running commentary that Π provides for this execution is actually a derivation in Π of the assertion that Program *ISHP-ALT never halts*, when given input *SI*. If that is true, then we have an *explicit* derivation in Π of Statement *IA*, found by the execution of Program *ISHP-ALT*, and we also see that there is a derivation in Π of Statement *IB* (coming from the running commentary). Hence in Case 2, as in Case 1, the formal system Π is *inconsistent*.

The subtle issue in Case 2 (that didn't arise in Case 1) is how the running commentary can conclude that Program *ISHP-ALT* never halts, because, as we established in our discussion of the Self-Halting Problem, there can be no program that correctly solves the Self-Halting Problem for all programs. The answer is that Program *ISHP-ALT* only fails to halt when it enters the Infinite loop detailed earlier. If the program does not enter the Infinite loop, it will eventually halt. We certainly can build into the program that creates the running commentary, the ability to recognize when the Infinite loop is entered, and to then conclude that Program *ISHP-ALT* will never halt.

[14] And also, not finding a derivation of Statement *IA*.

Cases 1 and 2 cover all possible executions of Program *ISHP-ALT*, and so Theorem 7.4.1 is proved. ■

What's the difference? The proof of Theorem 7.4.1 is almost the same as the second proof of Theorem 7.3.1. The two proofs have very similar case analyses, based on the possible executions of Program *SHP-ALT* and Program *ISHP-ALT*. But the two proofs differ in what is deduced from an execution.

In the second proof of Theorem 7.3.1, the difference between the *statement* (either *IA* or *IB*) that is explicitly generated by the execution of Program *ISHP-ALT*, and the *behavior* of Program *ISHP-ALT*, allows us to conclude that the program generated a *false* statement, and hence Π is *not sound*.

But in the proof of Theorem 7.4.1, the difference between the *statement* (*IA* or *IB*) that is explicitly generated by the execution of Program *ISHP-ALT*, and the *statement* that is derived by the running commentary based on the execution, allows us to conclude that the proof system Π derives *contradictory* statements, and is hence *inconsistent*.

7.5 *P* vs. *NP*: In the *Spirit* of Impossibility

Many readers may have heard about the *P vs. NP* problem, as it has great theoretical and practical importance and is widely taught and discussed in courses on algorithms and theoretical computer science. And, it is often presented in popular science venues as if it were an impossibility result. Maybe someday it will be – but it is not now! Still, it is in same *spirit* as impossibility, so I want to say a few words about it.

The *P vs. NP* problem concerns the *efficiency* of computational methods (algorithms) that solve *decidable* questions, that is, questions that *can* always be answered in finite time by computer programs. But, some of those questions might not be able to be solved *efficiently* (in a formally defined sense that we will not detail here). Questions that are decidable but can't be solved efficiently are often called *intractable*.

The class of questions *P* is the class of questions that *can* be solved by efficient (in a formal sense) algorithms on a computer with only a single processor; and the class *NP* is the class of questions that can be solved by efficient algorithms, *if* we use a computer that has a kind of unrestricted *parallel* computation and an unrestricted number of processors.[15] Every question in *P* is, by definition of the classes, also in *NP*. Symbolically, $P \subseteq NP$.

[15] There are other, equivalent, ways to define *NP*, but the one here is (I think) the simplest to state.

The *P vs. NP* problem is whether the opposite relation is also true, that is, whether $NP \subseteq P$. Equivalently, are all the questions that are in the class *NP* actually in the class *P*, or not? If *yes*, then the classes *P* and *NP* are the same class, that is, $P = NP$.

Another way to state the problem: Is unrestricted parallel computation necessary in order to efficiently solve all the questions in *NP*, or is a single processor always sufficient for efficient computation?[16]

The *P vs. NP* problem is currently *open* (unsolved), and there is an unclaimed million-dollar prize for solving it. Most theoretical computer scientists believe that the answer to the problem is *no*.[17] What is known is the following amazing theorem (yes, a really proved theorem): There is a class of questions in *NP*, which are not currently known to be in *P*, called the *NP-complete* questions, such that if *even one* *NP*-complete question is actually in *P*, then *every* question in *NP* is in *P*. That is, if even one *NP-complete* question has an efficient solution, then every question in *NP* does, and so the classes *P* and *NP* are actually the same class. A huge number of important questions are known to be in the *NP-complete* class.

So, if someday someone solves the *P vs. NP* problem by showing that the answer is *no*, then we would have a *proven* impossibility theorem: It would be proven impossible for even one *NP*-complete question to be solved by an efficient (in a formal sense) algorithm. But for now, the *P vs. NP* problem is open, and so we cannot state this as a known impossibility theorem. And for that reason, the *P vs. NP* problem really isn't a proper topic for this book, which is about impossibilities that have been *proven*. But, stay tuned – maybe in the second edition of this book, it will be a proper topic.

7.6 Exercises

1. We stated that when trying to prove that a problem is *undecidable*, it is logically correct to restrict attention to a subset of the instances of that problem. Explain this more fully.

   Is it also logically correct to restrict attention to a subset when we are trying to prove that a problem is *decidable*?

[16] An interesting historical note. In 1970, a letter written by Gödel to von Neumann in 1956 was discovered. This letter is now called *Gödel's Lost Letter* [64]. In it, Gödel posed the question of whether some decidable questions might be intractable, in the sense that they must take an impractically large number of steps (and hence time) to solve by a computer (well, actually he said "Turing machine"). So, in a way, Gödel anticipated the *P vs. NP* problem and the field of *computational complexity* that developed a few decades later.

[17] Most, but not all! Some famous theoreticians (e.g., Don Knuth) are open to the possibility that the answer is *yes*.

2. In proving that the Self-Halting problem is undecidable, we only needed to consider the entries on the main diagonal of Table *HP*. But in Chapter 6 when discussing our version of Gödel's theorem, we had to consider all the entries in Table *T*. Explain why we couldn't restrict attention to only the main diagonal in Chapter 6.
3. **Sharper undecidability** We have proved that the Self-Halting Problem is undecidable, so for any program, *SHP-PROP*, that is *proposed* to solve the Self-Halting Problem, there must be some instance, s_i, of the problem that Program *SHP-PROP* cannot correctly solve. Notice that we have not said whether that instance is one where Program P_i (the program specified by the string-encoding s_i) halts or one where P_i does not halt, when given input s_i.

   Now, suppose that we modify the proposed Program *SHP-PROP*, so that when it is given any input s_i it also, in parallel with whatever else it would do, *executes* a copy of Program P_i, with input s_i. If that execution of P_i halts, then the modified Program *SHP-PROP*, call it Program *SHP-PROP*$'$, can report *correctly* that the solution to the problem instance, s_i, is *Yes*. Therefore, Program *SHP-PROP*$'$ will correctly solve any problem instances where the correct solution is *Yes*.

   Use the above observation, and the fact that the Self-Halting problem is undecidable, to establish:

   Theorem 7.6.1 *It is impossible to create a program that correctly solves every instance of the Self-Halting Problem, even if the program does correctly solve all instances where the correct answer is "Yes".*

4. For any instance, s_i, of the Self-Halting Problem, exactly *one* of the two statements, *RA* or *RB* (defined in Section 7.3) must be *true*. Both statements can be formed in Π, but the proof of Theorem 7.3.1 established that *neither* statement can be derived in Π, so we have:

   Theorem 7.6.2 *In any sound formal system* Π *that can reason about programs, there is a statement (either RA or RB above) that is* true *and can be formed in* Π, *but is not derivable in* Π.

   Theorem 7.6.2 does not say which of the two statements, *RA* or *RB*, is the true one, but we can deduce from Theorem 7.6.1 that it must be statement *RB*.

   Deduce it.
5. The proof of Theorem 7.3.1 uses the Self-Halting Problem, but many other undecidable problems could be used instead. To state this more formally and generally, we define a set *X* to be *enumerable* if there is a computer

program that can generate and list the elements in X, guaranteeing that for any specific element x that is X, element x will be generated in some *finite* amount of time. For example, in the case of the Self-Halting Problem, X is the set of all string-encodings of programs that halt when given their string-encoding as input. We saw that this set X is enumerable. Any Self-Halting program, P_i, will be generated and listed in finite time.

We further define a set X to be *decidable* if there is a computer program that can take in any string, x, and determine (in finite time) whether or not x is in set X. Otherwise, set X is *un*decidable. We proved that when X is the set of programs that halt when given their string-encoding as input, X is undecidable.

Finally, for an arbitrary set X, a formal system can "reason about X" if it can form the statements "x is in X" and "x is not in X", where x is a string that might (or might not) be in X.

Show that an incompleteness theorem like 7.3.1 can be proved using *any* undecidable set X which is enumerable. The formal system is assumed to be able to reason about X.

6. In Section 7.3.1, we said that the assumption that a formal system can "reason about arithmetic" appears weaker than the assumption that a formal system can "reason about programs". Explain this.

8
Even More Devastating: Chaitin's Incompleteness Theorem

8.1 Perplexing and Devastating

In this chapter, we discuss a perplexing and amazing impossibility theorem, due to Gregory Chaitin [22], that is so sweeping, and yet so (relatively) simple to prove, that it almost feels like there must be some kind of cheating involved.[1] But no, it is for real.

Martin Davis (a leading expert on computation theory) called it a "dramatic extension of Gödel's incompleteness theorem" and of "devastating import" and "... a remarkable limitation on the power of mathematics as we know it".

8.2 Strings and Compression

Recall that a *string* is just a sequence of symbols, such as "this is a string", that is created using symbols from some alphabet, abstractly denoted $\mathcal{A}$.[2]

Some strings can be highly compressed, that is, represented by much smaller strings that can be used to reconstruct the original string. The reconstruction is created by following instructions in a computer program (or even

[1] What do I mean by cheating? Let's see an example of a cheating proof. There is a joke about *uninteresting integers*. The claim is that there are no uninteresting integers, and we can prove it! The proof goes like this:

Suppose there are uninteresting integers. Then, there must be a *smallest* uninteresting integer, call it z. Wow, you say that z is the smallest uninteresting integer? Boy, z is really special, and that makes it interesting. So, there is no smallest uninteresting integer, and it follows that there are no uninteresting integers. **Q.E.D.**

Have I actually proved that there are no uninteresting integers? Of course not – we have just played around with the words "interesting" and "uninteresting". The joke is a play on words, but also a play on the what a mathematical proof is. The structure of this "proof" is very similar to the structure of Chaitin's proof, although his is actually a proof, despite the similarity to this joke.

[2] We don't need to specify exactly what alphabet $\mathcal{A}$ is in order to state and prove Chaitin's theorem, so our use of "$\mathcal{A}$" is simply to denote some fixed, but unspecified, alphabet.

instructions in a natural language) that may have many fewer symbols than the string has. For example, the string

0101010101010101010101010101010101010101010101010101010101010101

has 63 symbols, but it can generated by the instructions:

REPEAT 32 times:
PRINT “01”

which has only 28 symbols (including spaces).

> **Review question:** There are actually instructions to print the above string using fewer than 28 symbols. Find such instructions. I think 19 symbols are enough.

And, if we want to make the compression even more impressive, consider the string that is generated by changing 32 to 32000000. That just adds six characters to the instructions but makes the generated string one-million times longer than before.

Note that these instructions are also strings, but instructions can use symbols from a different alphabet than $\mathcal{A}$. The alphabet used for instructions is abstractly denoted $\mathcal{A}_{\mathcal{L}}$. Alphabet $\mathcal{A}_{\mathcal{L}}$ will generally contain all symbols in $\mathcal{A}$, since instructions generate strings from the alphabet $\mathcal{A}$, and any symbol that appears in a generated string must be in alphabet $\mathcal{A}_{\mathcal{L}}$.

On the other hand Not all strings can be compressed in the way described above. If you have a string consisting of a seemingly random series of 70 bits (which I got from looking at the heads and tails of coins picked from a mixed-up box of coins), say

0011011001111010100111100011011001011001001110110001110

it is not obvious whether there are instructions that can generate this string, where the instructions have fewer symbols than the string itself. In fact, the instructions might need eight more symbols than the string has. It might be that the shortest instructions are:

```
PRINT "0011011001111010100111100011011001011001001110110001110"
```

As in the two examples above, we are concerned with compressing strings written in an alphabet denoted abstractly $\mathcal{A}$. Any such string, x, will be generated and output by *executing* instructions written in a fixed language (a chosen natural language, or a computer programming language), abstractly denoted

$\mathcal{L}$, whose alphabet is denoted $\mathcal{A}_\mathcal{L}$. We won't need to specify exactly what alphabet $\mathcal{A}_\mathcal{L}$ is.

Kolmogorov complexity We let $K_\mathcal{L}(x)$ denote the *smallest* number of symbols of any instructions, written in language $\mathcal{L}$, that generates and outputs string x, when the instructions are executed as a computer program. Note that the program generates string x without being given any input. Hence, each such program generates (at most) one string, and it generates the same string each time it is executed.

The number $K_\mathcal{L}(x)$ is called the *Kolmogorov complexity* of x (relative to language $\mathcal{L}$). The concept of Kolmogorov complexity was introduced by A. Kolmogorov in [58], but was independently discovered by others as well (e.g., G. Chaitin, R.J. Solomonoff).

Certainly, the value of $K_\mathcal{L}(x)$ depends on the language $\mathcal{L}$ used for instructions. Language $\mathcal{L}$ is assumed to be fixed, but we don't need to be more specific about what language $\mathcal{L}$ is. For our purposes, we only need to realize that once a language $\mathcal{L}$ is specified, there will be a definite value $K_\mathcal{L}(x)$ for every string x formed from alphabet $\mathcal{A}$. After we explain Chaitin's incompleteness theorem, we will see that the choice of language $\mathcal{L}$ is not critical to the theorem (it only changes a constant number in the theorem).

8.3 Chaitin's Incompleteness Theorem

Chaitin's incompleteness theorem[3] [22] concerns the provability of statements about the function $K_\mathcal{L}(x)$. It shows that there are statements about $K_\mathcal{L}(x)$ that *must be true* but *cannot* be derived in *any* consistent formal system Π. In that sense, it is similar to Gödel's first incompleteness theorem, but actually stronger.

The only thing we have assumed about Π is that it is consistent, but clearly Π must also have the ability to form (express) statements "$K_\mathcal{L}(x) > u$", for any string x and any number u, or else Chaitin's theorem will be trivially true for Π.

8.3.1 Statement of Chaitin's Theorem

Theorem 8.3.1 *Given an instruction language $\mathcal{L}$, there is a positive integer $u_\mathcal{L}$ such that for every string x formed from alphabet $\mathcal{A}$, the statement "$K_\mathcal{L}(x) > u_\mathcal{L}$" cannot be derived in any consistent formal system Π. Notice that $u_\mathcal{L}$ is a fixed number, independent of what string x is.*

[3] Which he proved while still in high school.

However, there are an infinite number of strings x, where $K_{\mathcal{L}}(x) > u_{\mathcal{L}}$. So, there are an infinite number of such true statements that can be formed in Π but cannot be derived in Π – the formal system Π is incomplete.

We will prove Chaitin's theorem, but we first show that the function $K_{\mathcal{L}}$ is *non-computable*. That proof is easier than the proof of Chaitin's theorem, and has several of the same elements, making it a good way to start.

8.4 $K_{\mathcal{L}}$ Is Not Computable

In Chapter 7, we saw a close connection between incompleteness and *non-computability*, specifically for the function $\overline{f}$. In this chapter, we have stated (and will prove) Chaitin's incompleteness theorem for the function $K_{\mathcal{L}}(x)$. So, it should not be a surprise that

Theorem 8.4.1 *Function $K_{\mathcal{L}}$ is non-computable.*

Proof For a proof by contradiction, suppose that the function $K_{\mathcal{L}}$ *is* computable by some program denoted $\mathcal{P}_{\mathcal{K}}$. That is, Program $\mathcal{P}_{\mathcal{K}}$ takes in a string x written in alphabet $\mathcal{A}$ and outputs the value $K_{\mathcal{L}}(x)$. We will show that if Program $\mathcal{P}_{\mathcal{K}}$ exists, then a contradiction arises, so we must conclude that $\mathcal{P}_{\mathcal{K}}$ does not exist – and that function $K_{\mathcal{L}}$ is non-computable.

Consider a program, called $P_{\mathcal{A}}$, that generates strings written in an alphabet $\mathcal{A}$, in order of their length (and within a specific length, it generates the strings in lexicographic order). Then, for each string x that $P_{\mathcal{A}}$ generates, the assumed program $\mathcal{P}_{\mathcal{K}}$ will compute the number $K_{\mathcal{L}}(x)$. Programs $P_{\mathcal{A}}$ and $\mathcal{P}_{\mathcal{K}}$ alternate execution, so that after $P_{\mathcal{A}}$ generates a string x, Program $\mathcal{P}_{\mathcal{K}}$ computes $K_{\mathcal{L}}(x)$ before $P_{\mathcal{A}}$ again generates another string. We combine programs $P_{\mathcal{A}}$ and $\mathcal{P}_{\mathcal{K}}$ in a program called $P_{\mathcal{AK}}$, which coordinates their executions, and also decides what to output and when to stop.

Now programs $P_{\mathcal{A}}$ and the assumed Program $\mathcal{P}_{\mathcal{K}}$ are finite-sized texts, and Program $P_{\mathcal{AK}}$ that combines them needs only a constant number of symbols in addition to those in $P_{\mathcal{A}}$ and $\mathcal{P}_{\mathcal{K}}$. So, Program $P_{\mathcal{AK}}$ only contains some constant number of symbols, denoted $c_{\mathcal{AK}}$, independent of whatever output it produces.

How does $P_{\mathcal{AK}}$ halt? We write Program $P_{\mathcal{AK}}$ so that it will halt *if and only if* Program $P_{\mathcal{A}}$ generates a string x where Program $\mathcal{P}_{\mathcal{K}}$ determines that $K_{\mathcal{L}}(x)$ is greater than the constant number $c_{\mathcal{AK}}$. See Figure 8.1.

Now, for every positive integer n, there are only a *finite* number of computer programs that contain n or fewer symbols, and each of those programs will

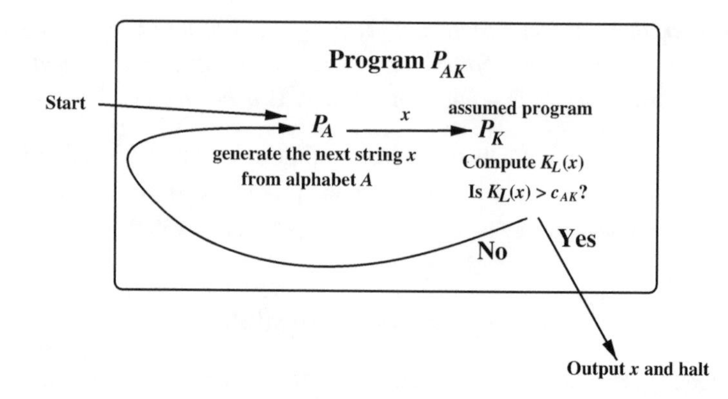

Figure 8.1 The conceptual Program $P_{\mathcal{AK}}$ used to prove that function $K_{\mathcal{L}}$ is not computable. Program $P_{\mathcal{AK}}$ has $c_{\mathcal{AK}}$ symbols. Program $P_{\mathcal{A}}$ produces strings from alphabet $\mathcal{A}$ in order of length (breaking ties lexicographically).

produce at most one string when executed without any input. So, there are only a finite number of strings that can be generated by programs that take no input and contain at most n symbols. That is, the size of the set $\{x|K_{\mathcal{L}}(x) \leq n\}$ is finite. In particular, there are only a finite number of strings that can be produced by such programs that have at most $c_{\mathcal{AK}}$ symbols. Note that we don't know, or need to know, what that finite number is.

Hence, Program $P_{\mathcal{A}}$ will ultimately (but in finite time) generate a first string x where $K_{\mathcal{L}}(x) > c_{\mathcal{AK}}$. At that point, the assumed Program $\mathcal{P}_{\mathcal{K}}$ will compute the number $K_{\mathcal{L}}(x)$, and then Program $P_{\mathcal{AK}}$ will output x and halt. But since Program $P_{\mathcal{AK}}$ only contains $c_{\mathcal{AK}}$ symbols, this would give explicit demonstration that $K_{\mathcal{L}}(x) \leq c_{\mathcal{AK}}$. That would be a contradiction to the assumption that Program $\mathcal{P}_{\mathcal{K}}$ correctly computes the function $K_{\mathcal{L}}$, and the fact that Program $P_{\mathcal{AK}}$ halts only when $\mathcal{P}_{\mathcal{K}}$ determines that $K_{\mathcal{L}}(x)$ is greater than $c_{\mathcal{AK}}$.

To summarize, the assumption made at the start of this proof, that Program $\mathcal{P}_{\mathcal{K}}$ exists, leads to a contradiction. So we must conclude that there can be no such program $\mathcal{P}_{\mathcal{K}}$ that computes the function $K_{\mathcal{L}}(x)$. Function $K_{\mathcal{L}}(x)$ is *non-computable.* ■

8.5 Now Back to Chaitin's Theorem and Proof

Our approach to proving Chaitin's theorem, stated in Section 8.3, is to first prove a version of the theorem where the word "consistent" is replaced by the word "sound". (That proof is similar, but differs in some critical ways, to the proof just given for Theorem 8.4.1.) Then, we will discuss how to modify that proof to prove Chaitin's theorem as stated above. But before we can present any proofs, we need to develop a few (four actually) *preparatory points.*

Preparatory point 1: More about instructions In Section 8.2, we saw two examples of instructions whose execution output strings where the instructions were "natural" and directly related to the strings they output. But, there is no requirement that instructions be natural or intuitive. The instructions can be tricky, unintuitive, or only *indirectly* related to the strings that are output. The only requirement is that when instructions are executed, a string x from alphabet $\mathcal{A}$ is output. Then, if those instructions have $S(x)$ symbols, this establishes that $K_{\mathcal{L}}(x) \leq S(x)$. The proof of Chaitin's theorem involves instructions that are (at first glance) only *indirectly* related to the strings they generate. So, be patient until you see Chaitin's clever idea.

Now, assume we have a formal proof system Π which is sound and can form statements $K_{\mathcal{L}}(x) > u$. The statements in Π are formed from an alphabet $\mathcal{A}_{\Pi}$.[4]

8.5.1 Preparatory point 2: Program P_{gc}

Consider a computer program, P_{gc}, in the specified computer language $\mathcal{L}$ consisting of two subprograms P_g and P_c, written in language $\mathcal{L}$. Program P_{gc} will also use the fixed positive integer $u_{\mathcal{L}}$ mentioned earlier. The actual value of $u_{\mathcal{L}}$ will be discussed later.

Program P_g successively *generates* strings written with the alphabet $\mathcal{A}_{\Pi}$, ordered by length, and for strings of the same length, ordered lexicographically. Each generated string, w, is a potential *derivation* in Π of a statement of the form "$K_{\mathcal{L}}(x) > u_{\mathcal{L}}$", for some string x (formed from alphabet $\mathcal{A}$). We have previously discussed similar programs, so it should be clear that Program P_g can be created.

The other Program, P_c, *checks* each string, w, generated by P_g, to see if w is a proper derivation of a statement "$K_{\mathcal{L}}(x) > u_{\mathcal{L}}$", for *some* string x. If it is, then the program outputs x, and P_{gc} stops. See Figure 8.2.

More precisely, what does the Program P_c need to check? The key is that string w is a potential derivation in Π of a statement "$K_{\mathcal{L}}(x) > u_{\mathcal{L}}$" for

[4] It is a bit confusing that we have three alphabets, $\mathcal{A}$ (the alphabet for strings), $\mathcal{A}_{\mathcal{L}}$ (the alphabet for instructions), and now $\mathcal{A}_{\Pi}$, the alphabet for the formal system Π. But, these alphabets are not independent. We have already noted that $\mathcal{A}_{\mathcal{L}}$ must contain $\mathcal{A}$, and we will see that $\mathcal{A}_{\mathcal{L}}$ also contains alphabet $\mathcal{A}_{\Pi}$. So, it's not as bad as it might seem at first.

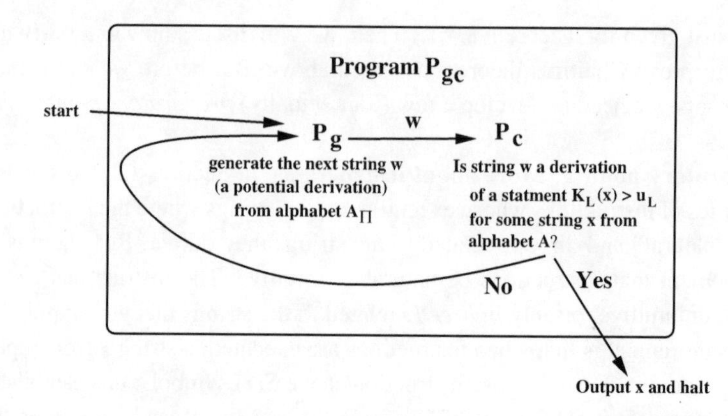

Figure 8.2 Program P_{gc} used to prove Chaitin's impossibility theorem. Programs P_g and P_c alternate execution.

some string x, but we do not specify what string x is before Program P_{gc} is executed. String x is generated as part of string w. So, P_c checks w to see that the first statements in w are axioms in the formal system Π, and then checks that each successive statement in w is obtained by applying some rule of logic (in Π) to an axiom or to a previously derived statement. Finally, P_c checks that w ends with the statement "$K_{\mathcal{L}}(x) > u_{\mathcal{L}}$", for some string x written with alphabet $\mathcal{A}$.

Program P_c is also similar to the ones discussed previously, and so it should be clear that Program P_c can be created.

8.5.1.1 Preparatory point 3: Programs P_g and P_c Alternate Executions

The executions of subprograms P_g and P_c alternate in Program P_{gc}, in the same way that subprograms in the previous two chapters alternate. So, Program P_{gc} successively generates strings formed from alphabet $\mathcal{A}_\Pi$ (where the strings are ordered by length, and within a given length, they are ordered lexicographically) and checks to see if a generated string, w, is a derivation of a statement "$K_{\mathcal{L}}(x) > u_{\mathcal{L}}$", for some string x. If it is, then P_{gc} outputs x, and stops. See Figure 8.2.

The key insight If Program P_{gc} outputs a string x, then we have an *explicit confirmation* that $K_{\mathcal{L}}(x) \leq |P_{gc}|$, where $|P_{gc}|$ denotes the number of symbols in Program P_{gc}. This is a bit subtle, because (unlike the two examples in Section 8.2), string x is not specified until Program P_{gc} is executed. Still, if the execution of P_{gc} results in the output of string x, that is a legitimate demonstration that $K_{\mathcal{L}}(x) \leq |P_{gc}|$, for the *specific* string x that is output. To see this, just recall the definition of $K_{\mathcal{L}}(x)$ and the fact that P_{gc} outputs string x. We can now prove an important claim.

Claim 1: Suppose there is a derivation, w, in Π, of the statement "$K_{\mathcal{L}}(x) > u_{\mathcal{L}}$" for *even one* string x. Then Program P_{gc} will, in finite time, generate a derivation, z, of a statement "$K_{\mathcal{L}}(y) > u_{\mathcal{L}}$", for some string y (which might be string x).

To establish this claim, note that since P_g successively generates strings written with alphabet $\mathcal{A}_\Pi$ in order of their length, there can only be a *finite* number of strings that might be generated and checked before string w would be generated. If Program P_{gc} does not stop before w is generated, then at that point Program P_c will recognize that w is a derivation of the statement "$K_{\mathcal{L}}(x) > u_{\mathcal{L}}$", and will output x and stop. So, there will be a *first*-generated derivation z (possibly w), of a statement of the form "$K_{\mathcal{L}}(y) > u_{\mathcal{L}}$", for some string y (possibly x). Program P_{gc} will then output y and stop. This proves Claim 1.

8.5.1.2 How Big Is Program P_{gc}?

Since Chaitin's theorem concerns Kolmogorov complexity, and Kolmogorov complexity concerns the number of symbols in a program that generates a string, we must determine the number of symbols in program P_{gc}. Program P_{gc} is formed from programs P_g and P_c, and some additional statements in P_{gc} needed to coordinate the alternating executions of those two programs. We first examine the number of symbols in P_g and P_c.

The number of symbols in Program P_g is some finite *constant*, call it c_g, which is *independent* of the value of $u_{\mathcal{L}}$. Program P_g simply enumerates strings written with alphabet $\mathcal{A}_\Pi$.[5]

Now note that *most* of Program P_c is also independent of the value of $u_{\mathcal{L}}$. Program P_c examines a string, w, produced by P_g, to see if w is a derivation of some statement "$K_{\mathcal{L}}(y) > u_{\mathcal{L}}$". Each string y is generated when Program P_{gc} is executed, and is not a "built-in" part of Program P_c. Hence, the length of y (no matter how large it gets) does not contribute to the number of symbols in P_{gc}. However, $u_{\mathcal{L}}$ is a fixed integer and *is* a built-in part of P_c. That is, the text making up Program P_c must explicitly contain the number $u_{\mathcal{L}}$. So, in addition to the parts of P_c that are independent of $u_{\mathcal{L}}$, Program P_c must explicitly represent number $u_{\mathcal{L}}$, and that representation must take some number of symbols. Note that we have not yet established what number $u_{\mathcal{L}}$ is. We will soon.

[5] If this point is unclear, try writing instructions in English to enumerate all the strings formed from $\mathcal{A}_\Pi$ in order of their length (and lexicographically for equal-length strings). Observe that those instructions use a finite number of symbols, independent of $u_{\mathcal{L}}$.

Now even though we don't yet know what number $u_{\mathcal{L}}$ is, we can ask how many symbols are required to represent it? Let's say that number $u_{\mathcal{L}}$ is written in normal base-ten notation, and use $di(u_{\mathcal{L}})$ to represent the number of digits needed to represent $u_{\mathcal{L}}$.[6] Then, the total number of symbols in Program P_c is $c_c + di(u_{\mathcal{L}})$, where c_c is a constant number equal to the number of symbols in P_c, other than the symbols (digits) used to represent $u_{\mathcal{L}}$.

Finally, some part of Program P_{gc} will coordinate the alternating executions of P_g and P_c, and it should be clear that this part of P_{gc} will only need a constant number of symbols, say c_p, independent of what $u_{\mathcal{L}}$ is. So all together the number of symbols in Program P_{gc}, denoted $|P_{gc}|$, is $c_g + c_c + c_p + di(u_{\mathcal{L}}) = c + di(u_{\mathcal{L}})$, for some constant number c. Hence,

Claim 2: If P_{gc} outputs a string y, we have an *explicit* demonstration that $K_{\mathcal{L}}(y) \leq |P_{gc}| \leq c + di(u_{\mathcal{L}})$.

8.5.1.3 Preparatory point 4: How to choose $u_{\mathcal{L}}$

Finally, how do we choose this mysterious number $u_{\mathcal{L}}$? As we will see, we want $u_{\mathcal{L}}$ to be *larger* than $c + di(u_{\mathcal{L}})$. But, is there always a value for $u_{\mathcal{L}}$ such that $u_{\mathcal{L}} > c + di(u_{\mathcal{L}})$, or equivalently, such that $u_{\mathcal{L}} - di(u_{\mathcal{L}}) > c$?[7]

To see that the answer is "yes", think about the following observations. The numbers {0...9} are each represented by one digit. So, ten numbers are represented by one digit, and the largest of those numbers is $10 - 1$. Then, the numbers {0...99} are represented by two digits. There are 100 of those, and the largest of those numbers is $100 - 1$. Next, the numbers {0...999} are represented by three digits. There are 1000 of those, and the largest of those is $1000 - 1$. We can continue in this way, seeing that at each step in this reasoning, each addition of one digit leads to the increase, by a multiple of 10, of the number of numbers that are represented, and that the largest number represented is always exactly one less than the number of numbers represented. More quantitatively, after d steps in this reasoning, the number of numbers represented by d digits is 10^d. We see that $10^d - d$ increases without bound, as d increases. So, no matter what the constant number c is, there will be some number of steps in this reasoning, call it d^*, where $10^{d^*} - d^* > c$, and $10^{d^*} - 1 - d^* \geq c$. Then, if we take one more step, we will have

[6] Many readers will know that $di(u_{\mathcal{L}}) = \lceil \log_{10} u_{\mathcal{L}} \rceil$, but I promised that everything in this book will be proven by pure reason alone – no need for any knowledge of mathematics beyond arithmetic and simple logic. So, in what is coming, we will not need to know that $di(u_{\mathcal{L}}) = \lceil \log_{10} u_{\mathcal{L}} \rceil$.

[7] For those readers who know that $di(u_{\mathcal{L}}) = \lceil \log_{10} u_{\mathcal{L}} \rceil$, we just need to see that there is a value of $u_{\mathcal{L}}$ such that $u_{\mathcal{L}} > c + \log_{10} u_{\mathcal{L}}$. Since c is a constant number, and $u_{\mathcal{L}}$ grows faster than $\log_{10} u_{\mathcal{L}}$ (in fact $u_{\mathcal{L}} / \log_{10} u_{\mathcal{L}}$ goes to infinity), there certainly will be a value of $u_{\mathcal{L}}$ that satisfies the inequality. If you followed this, you can skip the next paragraph.

$10^{d^*+1} - 1 - (d^* + 1) > c$. So, if we set the value of $u_{\mathcal{L}}$ to be $10^{d^*+1} - 1$, the number $u_{\mathcal{L}}$ can be represented with $di(u_{\mathcal{L}}) = d^* + 1$ digits, and for this value of $u_{\mathcal{L}}$, $u_{\mathcal{L}} - di(u_{\mathcal{L}}) > c$, as desired. In summary, once we know c (which is a constant based on P_{gc}) we can choose a value of $u_{\mathcal{L}}$ such that $c + di(u_{\mathcal{L}}) < u_{\mathcal{L}}$.

8.6 And Now the Proof

Having discussed four preparatory points, we can now prove Chaitin's theorem (the soundness version).

From Claim 1, we know that *if* there is a derivation, w, in the formal system Π, of a statement "$K_{\mathcal{L}}(x) \geq u_{\mathcal{L}}$", for even one string x, then Program P_{gc} will, in finite time, find a derivation, z, in Π, of a statement "$K_{\mathcal{L}}(y) > u_{\mathcal{L}}$" for some string y. It will then output string y, and stop.

But, from Claim 2, we know that $K_{\mathcal{L}}(y) \leq |P_{gc}| = c + di(u_{\mathcal{L}}) < u_{\mathcal{L}}$, and, we established that $c + di(u_{\mathcal{L}} < u_{\mathcal{L}}$, so $K_{\mathcal{L}}(y) < u_{\mathcal{L}}$. Hence, the supposed derivation z would be a derivation of a *false* statement. Since we assumed that Π is sound, no derivation in Π of a false statement is possible. That is, since Π is sound, if $K_{\mathcal{L}}(y) \leq u_{\mathcal{L}}$, the statement "$K_{\mathcal{L}}(y) > u_{\mathcal{L}}$" cannot be derived in Π. This leads to the conclusion that there cannot be a derivation in Π of the statement "$K_{\mathcal{L}}(x) > u_{\mathcal{L}}$", for *even one* string x formed from alphabet $\mathcal{A}$. This proves Chaitin's theorem (soundness version).■

8.6.1 Shocking?

Of course, Chaitin's theorem would not be shocking, or interesting, if $K_{\mathcal{L}}(x)$ is actually less than or equal to $u_{\mathcal{L}}$ for every string x formed from alphabet $\mathcal{A}$.

Yes, Shocking! Since the size of alphabet $\mathcal{A}_{\mathcal{L}}$ is finite, there can only be a finite number of strings formed from $\mathcal{A}_{\mathcal{L}}$, where the strings have length less than $u_{\mathcal{L}}$. Hence, there can only be a finite number of programs in the language $\mathcal{L}$ that have at most $u_{\mathcal{L}}$ symbols. By definition, the statement "$K_{\mathcal{L}}(x) \leq u_{\mathcal{L}}$" means that there must be a program in $\mathcal{L}$ of length less than or equal to $u_{\mathcal{L}}$ that outputs string x and no other strings. So, there can only be a finite number of strings x formed from alphabet $\mathcal{A}$ where $K_{\mathcal{L}}(x) \leq u_{\mathcal{L}}$. Hence, there must be an *infinite* number of strings x where $K_{\mathcal{L}}(x) > u_{\mathcal{L}}$.

John Baez put it this way:

> Lots of long strings of bits can't be compressed. You can't print out most of them using short programs, since there aren't enough short programs to go around. [8]

And "lots" really means "an infinite number".

In summary, we have proved: For any sound formal system Π that can form statements "$K_{\mathcal{L}}(x) > u$", there are an infinite number of strings, x, for which that statement is true, but *none* of those statements can be derived in Π. Hence, the formal system Π is incomplete. In fact, it is *extremely* incomplete.

8.6.2 Replacing "Sound" with "Consistent"

As stated earlier, Chaitin's statement of his incompleteness theorem assumes that the formal system is *consistent*, rather than *sound*, but we have only proved a version of the theorem where soundness is assumed. Soundness is a stronger (semantic) assumption, since any sound formal system must be consistent, but the converse need not be true. So, how did Chaitin prove his theorem with only the (purely syntactic) assumption of consistency?

Well actually, I am not going to be able to fully explain that, but I will explain the key insight. In the proof of Chaitin's theorem (semantic version), the only time we used the assumption that the formal system Π is sound was to observe (toward the end of the proof) that:

> ... if $K_{\mathcal{L}}(y) \leq u_{\mathcal{L}}$, the statement "$K_{\mathcal{L}}(y) > u_{\mathcal{L}}$" cannot be derived in Π.

Let us call this "The Key Observation". So, if we can also establish the Key Observation from the assumption that the system is *consistent*, rather than *sound*, then the above proof will work to prove the originally stated version of Chaitin's theorem.

Now, it is not immediate how the assumption of consistency would allow us to prove the Key Observation. The assumption of consistency only says that the formal system cannot derive the statement "$K_{\mathcal{L}}(y) > u_{\mathcal{L}}$" and also derive the statement "$K_{\mathcal{L}}(y) \leq u_{\mathcal{L}}$", which is not what the Key Observation says. But the function $K_{\mathcal{L}}(y)$ has a convenient property that we can exploit, namely:

> If $K_{\mathcal{L}}(y) \leq u_{\mathcal{L}}$, then there *is* a constructable, explicit demonstration of that fact – it is just the program (in language $\mathcal{L}$) with at most $u_{\mathcal{L}}$ symbols, whose execution outputs string y in a finite number of steps. Call this program $\mathcal{P}_y$.

This is the kind of property that is not true for every statement, but it is true here, and it is very helpful. With this property, a formal system Π can *derive* the statement "$K_{\mathcal{L}}(y) \leq u_{\mathcal{L}}$" if it can express Program $\mathcal{P}_y$ and then "execute" it (simulate its actions, step by step) to demonstrate that string y is output in finite time.

The following paraphrases (just converting some of Chaitin's terminology and notation to what we are using) the sentence in [22] on this point:

> ... if "$K_{\mathcal{L}}(y) > u_{\mathcal{L}}$" isn't true then there is a Program $\mathcal{P}_y$ that outputs y, and as this computation is finite, by carrying it out step by step in Π it can be proved that it works, and thus that $K_{\mathcal{L}}(y) \leq u_{\mathcal{L}}$.

This is the key (and subtle) point The expression, in Π, of Program $\mathcal{P}_y$ followed by a step-by-step simulation, in Π, of Program $\mathcal{P}_y$ with input y, which verifies that $\mathcal{P}_y$ does output string y, is a *derivation* in Π of the statement "$K_{LL}(y) \leq u_{\mathcal{L}}$". It might seem like a strange sort of derivation, but it actually is one.

Note that Program $\mathcal{P}_y$ does not have to be "discovered" through some derivation in Π; it only has to be expressed and executed in the formal system Π, since that provides a demonstration that the statement "$K_{\mathcal{L}} \leq u_{\mathcal{L}}$" *can be* derived in Π. Then, since Π is assumed to be consistent, the statement "$K_{\mathcal{L}}(y) > u_{\mathcal{L}}$" cannot also be derived in Π. This establishes the consistency variant of the Key Observation:

> If $K_{\mathcal{L}}(y) \leq u_{\mathcal{L}}$, then the statement "$K_{\mathcal{L}}(y) \leq u_{\mathcal{L}}$" can be derived in Π, and since Π is consistent, the statement "$K_{\mathcal{L}}(y) > u_{\mathcal{L}}$" cannot be derived in Π.

And, as stated earlier, once we have this version of the Key Observation, the original (consistency) version of Chaitin's theorem is established.

8.6.2.1 What Is Missing?

I said earlier that I would not be able to provide a full proof of the consistency version of Chaitin's theorem. Why not? Because I did not justify that the execution of $\mathcal{P}_y$ could be *simulated* (step by step) in Π. How is it that Π can simulate an execution of Program $\mathcal{P}_y$? Equivalently, considering the earlier quote from [22], what exactly does "carrying it out step by step in Π" really mean?

The ability of a formal system to do these kinds of simulations of the execution of a computer program (or of a more primitive model called a Turing Machine) is absolutely central in many proofs of incompleteness-related theorems. The key to these simulations is that the workings of a computer program can be re-expressed as statements about integer arithmetic. The amazing fact is that if the axioms of a formal system include a few (six is enough) simple statements about arithmetic on positive integers (such as the fact that zero is not the successor of any positive integer; or that every positive integer has a successor; or that for any two positive integers, p and q, $p + q = q + p$), then derivations in the formal system can simulate the step-by-step executions of a computer program. The details are *really ugly*, and are a large part of what Gödel invented in the proofs of his incompleteness theorems. In advanced courses on mathematical logic, students are usually tortured with working through some of the more tedious details,[8] but in research publications, the details are mostly avoided with the assertion that the formal system contains "a certain amount of arithmetic" or "can carry out a sufficient amount of arithmetic" or simply that it is "rich-enough". Readers in the intended audience then know that the needed simulations via arithmetic are possible in the intended formal system.

From this perspective, the literal statement of Chaitin's incompleteness theorem (the consistency version) in [22] is actually deficient, for it also should have stated that the formal system needs to be able to carry out a "sufficient" amount of arithmetic. Later, the paper sort of makes up for that omission by saying that the result holds for formal systems "unless the methods of deduction of the theory [formal system] are extremely weak."

8.7 What Is So "Devastating" about Chaitin's Theorem

Why is Chaitin's theorem considered of "devastating import" and "remarkable limitation on the power of mathematics as we know it", particularly coming well after Gödel's theorem, which already established limitations on the power of "axiom-based, provable" mathematics? Gödel's theorem showed that there are true statements that cannot be derived (proved) in any of the formal systems that are used in typical mathematical work.

Gödel's original proof[9] is based on *self-referential* paradoxes: "This statement is false", and "This statement cannot be proved". A "solution" to such paradoxes means making sense of those statements – but there are no satisfying solutions – which is why they are paradoxes.

Gödel's original proof showed how paradoxes like these can be encoded into statements about integers and arithmetic. So, head-spinning attempts to make sense of the paradoxes became head-spinning attempts to make sense of the translated statements about arithmetic. As shocking as Gödel's theorem was originally, eventually it seemed somewhat natural and acceptable that these paradox-based statements about integers could not be "proved", any more than the originating paradoxes could be "solved".

It seemed right that when you construct convoluted, self-referencing, knotted statements based on paradoxes that only a logician could love, they will have some defects – like not being able to be proved. In the end, I think that many mathematicians had the attitude:

[8] I was.

[9] Different from the type of proof we discussed in Chapter 7.

> OK, OK, Gödel showed that there are some unnatural, extreme statements that are unprovable, yet true,[10] but they are rare, weird, and convoluted things, precisely engineered to make Gödel's proof work. Those statements *deserve* their pariah status! It's no wonder that Gödel went crazy.[11] I will just stay over here in my safe space, in the part of math that makes sense. None of that Gödel stuff affects me. Or, as stated in [37]: "Gödel's Theorem, for most working mathematicians, is like a sign warning us away from logical terrain we'd never visit anyway."[12]

But then, Kolmogorov and Chaitin came along, turning attention to sensible statements about string compression. String compression is big business and hugely important in many technologies: communication, music, video, photography, DNA databases, etc. Methods BZIP, GZIP, LZW, JPEG, MP3,4, and 5 are just a few of the examples of string compression methods in wide use. So, there are *natural* statements about string compression that we *really* do want to make and prove. These statements aren't oddities.

Unlike the statement "This statement is false", which is a head spinner that would never come up in polite conversation, statements of the form "The string x cannot be generated by a program with fewer than u symbols" are prosaic, straightforward, and clear. It cleanly asserts a concrete bound on how much string x can be compressed into a program that re-creates x. That assertion may be true or it may be false, but what it means is not a riddle or a paradox.

Moreover, if the assertion is false, there is an easy proof of it – just display the program that generates x, and verify that it has fewer than u symbols. (Such a program may be hard to find, but it is easy to verify.) Contrast that with the situation of the statement "This statement is false". What would it even mean to prove that statement is false? It smells like the cheating argument about "uninteresting" integers.

And another thing. Most (in fact, an infinite number) of the statements "$K_{\mathcal{L}}(x) > u$", are not only true, they are *extremely* true. Most of these statements are weak, penny-ante, and pathetic. u is a constant number, while the length of x is unbounded. I read somewhere that for an axiom system that is sufficient for most of what mathematicians actually do, and using the programming language Python, u is around 2,000. Now, consider a string x that looks random and contains one million–trillion characters. The claim that all programs in Python that generate this one million–trillion-length random-looking string must contain at least 2,000 symbols seems pretty measly, trivial, and true.

[10] This is actually an incorrect statement.

[11] Gödel starved himself to death, convinced that he was being poisoned.

[12] Actually, this logical terrain is much larger, and is visited more often, than is implied by this statement. We will return to this issue at the end of the next chapter.

And yet, you mean to tell me that this pitiful, trifling, two-bit claim *cannot* be proved in any consistent logical system. That's embarrassing – and frightening.

In conclusion Gödel's proof of incompleteness and non-provability concerns a *few* highly engineered, show-offy, paradoxical statements that went out of their way to annoy. They "asserted ... weird and twisty things about themselves" and were "bizarrely twisty formulas" [52]. They show no remorse for their non-provability, and in my opinion were kind of "asking for it".

In contrast, Chaitin's proof of non-provability concerns an *infinite number* of true, almost obvious, innocuous statements that clearly never deserved to end up in such a mess. That is very disturbing. And, if this can happen to such nice, clean-cut, next-door-neighbor type statements, then it might happen to *your* favorite unproved statement. It is this frightful possibility that is of "devastating import", suggesting that "mathematics as we know it" will be forever limited.

8.7.1 OK, That Was Fun, but Seriously Now

I really enjoyed writing the previous section.[13] But seriously, if Chaitin's theorem is so much more natural, powerful, and devastating than Gödel's theorems, and is certainly easier to prove and explain, why has it not supplanted Gödel's theorems in logic and computer science courses, and popular press? Why is there still so much focus on Gödel?

Part of the answer is the historical importance of Gödel's theorems, the importance of being first, the huge impact that those results had, and the fact that they established Gödel the greatest logician (by far) of the twentieth century. Also, Chaitin's theorem is more related to Gödel's First incompleteness theorem than it is to Gödel's Second theorem. And, in order to discuss Gödel's Second theorem, one has to first discuss his First theorem and Gödel's original proof of it. But, I think there is another, more mathematical, reason as well.

Gödel's proofs established tools and techniques that have led to other incompleteness and impossibility theorems. They have also led to more general, abstract theorems and insights about incompleteness that have application beyond Gödel's primary focus. These generalizations more clearly explain why and how incompleteness arises. On the other hand, Chaitin's theorem and proof, as amazing as they are, seem in comparison to be very special, very

[13] Of course, I will edit it out before the book is published.

tied to Kolmogorov complexity. Chaitin's incompleteness does not seem as generalizable as Gödel's theorems are, or as widely applicable.

So, I think the continued importance of Gödel's theorems and proofs partly come from their greater generalizability. Consistent with that, in Chapter 9, we will discuss and prove an *abstract* version of Gödel's First incompleteness theorem that essentially gives a *recipe* for creating incompleteness results.

9
Gödel (For Real, This Time)

You *want* the truth? ... You *can* handle the truth![1]

Recall that in Chapter 6 we proved a Gödel-ish theorem that establishes the *popular statement* of Gödel's First incompleteness theorem (at the start of that chapter), but is weaker than Gödel's actual First incompleteness theorem in several respects. Then, in Chapters 7 and 8, we established other, related, incompleteness and undecidability results that also imply Gödel-ish theorems that are consistent with the popular statement. But, we didn't prove any actual theorem that came from Gödel himself.

In this chapter,[2] we will prove an *abstract* version of Gödel's First incompleteness theorem that is included in Gödel's original 1931 paper [45].[3] The abstract version is attractive because it cleanly encapsulates the high-level *logical* aspects of Gödel's theorem and Gödel's proof, without diving into the grunge needed to apply it to any particular formal system. It essentially gives a high-level roadmap of how to prove incompleteness (when possible) in many different logical systems.

After we prove the abstract theorem, we will give a grunge-free *sketch* of how it can be used to fully prove Gödel's First incompleteness theorem, once a formal proof system is specified. Our discussion of the abstract theorem will also provide a segue into our discussion and sketch of Gödel's *Second* incompleteness theorem.[4] We end the chapter by discussing some common

[1] Something that Jack Nicholson never said to Tom Cruise in the movie *A Few Good Men*.

[2] The exposition in this chapter is guided by the first two chapters of [100].

[3] At least, Smullyan [100] wrote that it is in Gödel's paper, and who am I to doubt Smullyan?

[4] The material presented in the technical sections of this chapter, Sections 9.3 through 9.6, is probably the most difficult material in the book. Chapters 6, 7, and 8 gave simpler, yet rigorous, proofs of incompleteness in contexts somewhat different from that of Gödel's

(technical) misconceptions about Gödel's two incompleteness theorems and about the causes of incompleteness.

9.1 Background

Recall from Chapter 6 that a *formal system*, denoted Π, consists of a (finite) alphabet and a language that uses the alphabet, a (finite) set of axioms, and a (finite) set of inference rules. The formal system is used to *derive* statements (written in the language of Π) by starting from the axioms and successively applying rules of inference. Any statement obtained in this way is said to be *derived* in Π.

Recall also that Gödel's main focus was on formal systems for *arithmetic*. So, while we will not specify any particular formal system Π, we will restrict the language of Π to be a *language of arithmetic* (defined shortly), denoted $\mathcal{L}_A$.

Then, we can abstractly refer to any formal system Π that contains $\mathcal{L}_A$ and develop Gödel's theorem in terms of that abstract Π.

A general statement again I think it's helpful to give another high-level statement of Gödel's theorems. The following combines both syntactic versions of Gödel's First and Second incompleteness theorems into a single statement:

> Today the incompleteness theorem is often formulated as a theorem about any formal system within which a certain amount of elementary arithmetic can be expressed and some basic rules of arithmetic can be proved. The theorem states that any such system, if consistent, is incomplete, and the consistency of the system cannot be proved within the system itself. [39]

What do we mean by "number" and "arithmetic"?[5]

Before discussing anything else, let me state clearly that in all that follows we are concerned with *non-negative integers*, {0, 1, 2, 3, ... }, which are also called *natural numbers*. So, whenever we use the word *number* or *integer*, the intended meaning is a *non-negative integer*. Now what do we mean by *"arithmetic"*? It is more than what you learned in grade school. By arithmetic, we mean statements and facts about *non-negative integers*. The term *arithmetic* in the sense discussed here is also called *number theory*.

primary concern. Those incompleteness theorems are also deep and profound, and (I think) have much of the philosophical impact of Gödel's actual First incompleteness theorem. But, for readers who really want to get into what Gödel actually established, and also in order to discuss Gödel's *Second* incompleteness theorem, this chapter is where we do it.

[5] A note on the use and omission of quote marks. I don't understand when to use them. So, I generally will not use quote marks, unless it is obvious that they are needed, or when there might be real ambiguity if they were omitted, or to indicate that a concept (such as "number", "arithmetic", or "truth") is a bit vague.

9.2 The Elements of $\mathcal{L}_A$, a Language of Arithmetic

The specific formal language we use in this discussion is denoted $\mathcal{L}_A$. It is a language that can be used to write and derive expressions (statements) concerning integer arithmetic – it is "*a* language of arithmetic".[6]

9.2.1 The Alphabet of $\mathcal{L}_A$

The alphabet we will use for language $\mathcal{L}_A$ will have 26 (what a coincidence!) symbols, separated by commas:

$$0, 1, 2, 3, 4, 5, 6, 7, 8, 9, +, \times, =, \leq, (,), x, \;', \wedge, \vee, \neg, \Rightarrow, \Leftrightarrow, \exists, \forall, \&$$

Review question: Does this alphabet contain a comma?

Familiar symbols The digits 0 through 9, and the symbols "+ (plus), × (times), =, ≤, (,)", need no explanation, as they have their normal meaning in arithmetic.[7]

Variables The symbol x in the alphabet of $\mathcal{L}_A$ denotes a single *variable*, which can be assigned any integer value. In forming expressions in $\mathcal{L}_A$, we will often need more than one variable, and there is no limit on how many we might need. Additional variables can be *created* in $\mathcal{L}_A$ by affixing one or more copies of the *prime* symbol " ′ " to the symbol x. For example, two such variables are x' and x'''.

An important notational convention The symbol x represents a variable in $\mathcal{L}_A$ that can be, but has *not* been, assigned a value. However, we sometimes need to represent the fact that x *has* been assigned a value, but we don't need to know what that value is. In that case, when we don't know or care what value x has been assigned, but we do care *that* it has been assigned a value, we

[6] In this chapter we use both words "expression" and "statement". These words are almost interchangeable. But there is often a subtle difference. The word "expression" is used to refer to anything that can be written in $\mathcal{L}_A$, while the word "statement" generally means an expression that has an actual meaning. This is vague (sorry), and you will not go far wrong if you think of the two words as synonymous.

[7] Typically in mathematical logic, the alphabet only has a single digit, 0, but would also have another symbol, S, which has the meaning: *successor of*. Then, for example, with that alphabet, we would write $SSS0$ to denote the number 3. There are some advantages to this convention, but they will not be needed in this exposition, so I see no reason to use such notation.

use the notation $\hat{x}$. So, for example, $E(x)$ is a function of variable x; but $E(\hat{x})$ is function E *evaluated* at a value $\hat{x}$, although we (in the exposition) have not specified what value has been assigned to x. Note that $\hat{x}$ is *not* a variable, and that the *hat* in $\hat{x}$ is not part of the alphabet of $\mathcal{L}_A$ – it is just a *reminder* to us that x has been given a value.

We will also use symbol n, which is a variable that can be assigned an integer, but it is *not* a variable in the language $\mathcal{L}_A$. It is a mathematical variable that we use in *our* thinking and in our proofs. As mentioned above, when we want to indicate that it has been assigned a specific integer value (that we don't need to specify), we use the notation $\hat{n}$.

The use of the *hat* symbol in $\hat{x}$ and $\hat{n}$ is meant as a *reminder* that, at that point in the exposition, the hatted variables have been *assigned* a fixed value. That convention is not universal, and if it causes you confusion, just ignore the hats (but then you will have to know yourself whether the symbol, x or n, denotes a variable, or a value, at that point in the exposition).

Logical symbols The symbols $\wedge$ and $\vee$ stand for the logical AND and OR operations.

Suppose X and Y are expressions that each already have a value *true* or *false*. Then the expression $X \wedge Y$ will have value *true* if and only if both X and Y have the values *true*. Expression $X \vee Y$ will have value *true* if and only if at least one of X or Y has the value *true*.

The symbol "$\neg$" stands for NOT. If expression X already has a value of *true* or *false*, then the expression $\neg X$ will have the value that is the *negation* of the value of X. So, $\neg X$ will have value *true* if and only if X has the value *false*.

Review question: Notice that the symbol "$\leq$" means "less than or equal", and that the alphabet for $\mathcal{L}_A$ does not contain the symbol "<", which means "strictly less than". Using only the symbols in the alphabet for $\mathcal{L}_A$ write an expression that means "$x < 5$".

Symbol "$\Rightarrow$" is shorthand for *IF – THEN*. For two statements X and Y, the expression "$X \Rightarrow Y$" means IF X THEN Y, or equivalently, X *implies* Y.

If X and Y already have values of *true* or *false*, then the expression $X \Rightarrow Y$ will have value *true*, *unless* X is *true* and Y is *false*. This is sometimes confusing because it means that when X is *false*, $X \Rightarrow Y$ will be *true* (no matter whether Y is *true* or *false*).

The related symbol "$\Leftrightarrow$" is shorthand for *IF AND ONLY IF*. That is, for two statements X and Y, $X \Leftrightarrow Y$ means "X if and only if Y". When X and Y already have values of *true* or *false*, the expression $X \Leftrightarrow Y$ will have value *true* exactly when the values of X and Y *agree*, that is, either both are *true* or both are *false*.

The expression $X \Leftrightarrow Y$ is also equivalent to: $(X \Rightarrow Y) \wedge (Y \Rightarrow X)$.

> **Review question:** If you have learned about *truth tables* in a logic or computer science or math course, verify that what has been stated above about the meanings of $\Rightarrow$ and $\Leftrightarrow$ agree with what you learned about truth tables.

Quantifiers The symbol "$\forall$" is the standard symbol used in mathematics as shorthand for the phrase "for all". It is usually followed by the name of a variable, say x, as in $\forall x$, which translates to English as "for all values of variable x". In this situation, we say that variable x is *bound* to the symbol $\forall$.

The related symbol "$\exists$" is the standard mathematical symbol used as shorthand for the phrase "there exists". It is also usually followed by a variable, as in $\exists x$, which translates to "there exists a value of variable x". Here, we say that x is *bound* to the symbol $\exists$.

End of expression The final symbol in the alphabet, "&", is used to signal the end of an expression. We will see the need for such an end-of-expression symbol when we discuss *encoding* and *decoding* of expressions and derivations.

9.2.2 Examples of Expressions in $\mathcal{L}_A$

Constants The simplest expression is a single constant. An expression is a called a *constant* if it has no variables, so that it unambiguously identifies a *unique* number. Examples of constants are: 15 (duh!), $(15 + 66)$, $((9 \times (6 + 3)) \times 7)$.

> **Review question:** Verify that the last two expressions identify the constant numbers 81 and 567.

Expressions with variables Of course, for useful expressions, we need more than constants. Consider the following expression:

$$\forall x \exists x' (x' = 2 \times x).$$

In this expression, x and x' are variables. Variable x is *bound* to the *for-all* symbol $\forall$, and variable x' is bound to the *there exists* symbol $\exists$. In English the expression says:

For any integer value assigned to variable x, there is an integer value of x' which is two times the number assigned to x.

This expression makes an assertion that is *true* in our normal, standard interpretation (i.e., understanding) of integer arithmetic. As a second example, consider:

$$\forall x \exists x'(x = 2 \times x').$$

Variables x and x' are again bound to the symbols $\forall$ and $\exists$, respectively, but this expression asserts that for any integer x, there is another integer x' such that x is two times x'. In other words, it asserts that every integer is an *even* number. Of course, that assertion is *false* in the normal interpretation of integer arithmetic.

Free variables and sentences A variable in an expression is said to be *free* if it is *not bound* to either a $\forall$ symbol or a $\exists$ symbol. For example, in the expression

$$\forall x(x' = 2 \times x),$$

the variable x' is free, while the variable x is not free – it is bound to the $\forall$ symbol.

An expression with *no* free variables is called a *logical sentence*. When there is no ambiguity, we will simply refer to a logical sentence as a *sentence*. For example, two previous expressions we looked at, $\forall x \exists x'(x' = 2 \times x)$ and $\forall x \exists x'(x = 2 \times x')$, are both sentences because they have no free variables.

The importance of distinguishing sentences from other expressions is that a sentence makes an assertion that is either *true* or *false*. That is, the meaning of the sentence conveys either a truth or falsehood. It may be difficult to determine whether a particular sentence is true or false, but it must be one or the other. Truth in $\mathcal{L}_A$ is defined by the normal interpretation of integer arithmetic, and so the first sentence above is true, while the second sentence is false in $\mathcal{L}_A$.

Review question: Which of the following four expressions are sentences: $x \leq 7$, $\exists x(x \leq 7)$, $\forall x \exists x'(x + x' \leq 7)$, $\forall x(x + x' \leq 7)$.

Making sentences Consider the expression:

$$\exists x(x = 10 - x'),$$

which is *not* a sentence since x' is a free variable. Clearly, it is not meaningful to assert that this expression is true or that it is false. Without knowing a value for x', the expression cannot be said to be true nor false.

However, if we *assign* an integer value to variable x', then the expression becomes a sentence, and that sentence *will* either be true or false.[8] For example, if we set the value of x' to 7, then the expression becomes the sentence:

$$\exists x(x = 10 - 7),$$

which is *true*.

Review question: What values, when assigned to x', would turn the expression $\exists x(x = 10 - x')$ into a true sentence? Recall that values of variables are only allowed to be non-negative integers.

9.2.3 Predicates

An expression with exactly *one free* variable, say x, which becomes a sentence when x is assigned an integer value, is called a *single-variable predicate*, with variable x.[9] For example, the previous expression $\exists x(x = 10 - x')$ is a predicate with free variable x'.

Naming predicates In the predicate

$$\exists x'(x = 2 \times x'),$$

the variable x is free. This predicate becomes a sentence when a value is assigned to x. That sentence is true (in the normal interpretation of integer arithmetic) if and only if an *even* number is assigned to x. Hence, it is natural to *refer* to this predicate as *EVEN(x)*.

As in this example, we often denote a predicate with uppercase letters, followed by the free variable in parentheses; and when possible, we give the predicate a name that suggests its meaning.

[8] For an analogy, consider the statement in English: "That woody plant is actually a tree." We can't determine whether this statement is true or false (in the normal interpretation of English) because we don't know which woody plant is being referred to. But if the statement is expanded to "The woody plant with ID tag number 731137 in the Chicago Botanical Gardens is actually a tree", then the plant is specified, and the sentence is either true (if the specified plant is a tree) or false (if the specified plant is not a tree).

[9] When we refer to a "predicate", that is shorthand for "single-variable predicate", unless stated otherwise.

Review question: Write out a predicate that might naturally be referred to as *ODD(x)*.

Predicates express (define) sets and properties In the above example, it is natural to say that the predicate *EVEN(x) defines* or *names* or *expresses*, in $\mathcal{L}_A$, the *set* of *even* integers. That is, the predicate *EVEN(x)* becomes a true sentence if and only if variable x is replaced by an even number.

Further, it is natural to say that *EVEN(x)* expresses, in $\mathcal{L}_A$, the *property* of being an even number. We also say that the property of being an even number is *expressible* in $\mathcal{L}_A$, precisely because we can create such a predicate in $\mathcal{L}_A$. Note that "express" does not mean "derive", as explained in the following quote:

> ... what a theory can *express* depends on the richness of the language (the definition doesn't mention proofs [derivations] or theorems) [99].

A more challenging example Recall that a number is *prime* if and only if it cannot be evenly divided by any number except *one* and itself. Equivalently, a number x is prime if and only if x is larger than *one*, and whenever it is the product of two numbers, x' and x'', at least one of x' or x'' *must* be the number *one*. Notice that in this definition, the number *one* is not considered to be prime.[10]

For example, the number 7 is prime, since the only way that 7 is the product of two numbers is if one of those numbers is *one*. However, the number 20 is not a prime, since 20 is the product of 4 and 5. In fact, no even number is prime.

Now, how do we express, in $\mathcal{L}_A$, the property that number 7 is prime? Consider the following expression:

$$(2 \leq 7) \wedge (\forall x' \forall x''((x' \times x'' = 7) \Rightarrow (x' = 1 \vee x'' = 1))).$$

What this says is that 7 is larger than *one*, and that for any two numbers x' and x'', if x' times x'' equals 7, then either the value of x' is *one* or the value of x'' is *one*. This is an exact translation of the second way we defined a prime number.

Now let's look at the situation where we don't have a constant number, such as 7, and we want to express the requirement that a variable number, denoted x, must be a prime. How do we do that? Consider:

[10] This is the more "modern" definition of a prime number. In the good old days, the number one was a perfectly fine prime, and I still secretly think it is. I have no idea why or when number one was banished, or why Pluto is not still a planet.

$$(2 \leq x)) \wedge (\forall x' \forall x''((x' \times x'' = x) \Rightarrow (x' = 1 \vee x'' = 1))).$$

In this expression, the variable x is free, but the other two variables are bound, so this expression is a *predicate*, with variable x. We refer to it as *PRIME(x)*.

Review question: Write the statement that there are an *infinite* number of prime numbers, a fact that was known since antiquity.

Hint: The assertion that there is an infinite number of prime numbers is equivalent to the assertion that there is no largest prime, that is, for every prime number, there is another prime number that is larger.

Remember that when we write *PRIME(x)*, that is just an abbreviation for the entire expression shown above. You can use it to avoid having to write out the full expression for *PRIME(x)*.

Goldbach As our final example, we will see how to express *Goldbach's conjecture* (introduced in Chapter 7) in $\mathcal{L}_A$. Recall that Goldbach's conjecture is that *every even* number larger than two is the sum of two prime numbers (possibly the same prime). Goldbach's conjecture may be hard to *resolve*, but it is easy to *express* in $\mathcal{L}_A$:

$$\forall x \exists x' \exists x'' \exists x'''(((2 < x) \wedge (x = 2 \times x')) \Rightarrow (PRIME(x'') \wedge PRIME(x''') \\ \wedge (x = x'' + x'''))).$$

And by dropping the initial "$\forall x$", we can form the predicate *GOLDBACH(x)*:

$$\exists x' \exists x'' \exists x'''(((2 < x) \wedge (x = 2 \times x')) \Rightarrow (PRIME(x'') \wedge PRIME(x''') \\ \wedge (x = x'' + x'''))),$$

which is true if and only if the value of x is an even number larger than two and is the sum of two prime numbers. So, predicate *GOLDBACH(x) expresses* the property of being in the set of numbers that *agree* with the Goldbach conjecture. We call those *Goldbach numbers*. Currently, it is unknown whether the Goldbach conjecture is true or false.

9.2.4 Formalizing the Concept of "Expressing"

Above, we used the terms "express", "expressible", and "expressing" for properties of sets of numbers. The general meanings of these terms may already be clear from these examples, but let's make their meaning fully explicit:

A *set* of numbers, say B, is *expressible* or *expressed* in $\mathcal{L}_A$ if there is a single-variable predicate, $E(x)$, written in $\mathcal{L}_A$, such that:

For any integer value $\hat{x}$ assigned to the variable x, the expression $E(\hat{x})$ is true if and only if $\hat{x}$ is in set B.

We also say that "B is expressed in $\mathcal{L}_A$ by predicate E".

Review question: What set of numbers is expressed by the predicate $(x \leq 15) \wedge PRIME(x)$?

Write a predicate in $\mathcal{L}_A$ that expresses the set of all even integers larger than or equal to 100.

9.2.5 Recapping

With the normal interpretation of integer arithmetic, we have seen both true and false sentences that can be written in $\mathcal{L}_A$, a language of arithmetic. And, we have seen a sentence that can be written in $\mathcal{L}_A$ where the truth of that sentence is currently unknown. We have also seen how these sentences are related to predicates that express the property of being in a specific set of numbers (e.g., the even numbers, the prime numbers, the Goldbach numbers).

9.2.6 An *Abstract* Formal System Π

Recall that any *formal system*, Π, has an alphabet, a language, a set of axioms, and rules of inference.

In this chapter, we have specified the elements of $\mathcal{L}_A$, a language of arithmetic, but we have said nothing about axioms or rules of inference of any specific formal system Π. Still, even without fully specifying a particular formal system, Π, we can abstractly conceive of, and reason about, the set of sentences that are *derivable* in a formal system.

In what follows, we use Π to refer *abstractly* to a formal system that contains the language $\mathcal{L}_A$. But, remember that $\mathcal{L}_A$ contains symbols for logical operations, and the meaning of these operations was also defined for $\mathcal{L}_A$. So, in this way, the rules of inference in our abstract Π are not completely undefined. Those rules must be consistent with the meanings of the logical operators in $\mathcal{L}_A$. No additional detail about Π is required in order to prove the abstract incompleteness theorem to come.

9.3 Gödel Numbers and the Diagonal Function

In order to get to a precise statement and proof of the abstract Gödel theorem we will prove, we need to develop two critical tools: *Gödel numbering* and *the diagonal function*.

9.3.1 First Key Tool: Gödel Numbers

We will need to translate expressions in $\mathcal{L}_A$ into numbers since predicates in $\mathcal{L}_A$ only define properties of sets of *numbers*. This translation also allows expressions in $\mathcal{L}_A$ to refer to expressions in $\mathcal{L}_A$, including *themselves*. The translation is done through a technical step called *Gödel numbering*. *Gödel numbering* is simply a way to encode any expression (in $\mathcal{L}_A$), denoted E, as an integer $g(E)$ (called the *Gödel number* of expression E), so that E can be referred to by its Gödel number $g(E)$.

Of course, *distinct* expressions must get *distinct* (i.e., different) Gödel numbers, so that any expression can be *unambiguously* reconstructed from its Gödel number. Any unambiguous encoding system where the encoding and decoding can be done *mechanically* (e.g., via a computer program) will work as a Gödel numbering.

Gödel invented a very intricate way to form Gödel numbers, involving powers of successive prime numbers. He needed a number scheme with properties that we don't need. In effect, he needed a number scheme that allowed him to "program without programming" [1], because computer programming had not been invented yet.[11] But, we (after Turing) do know about programming, so our scheme for forming Gödel numbers can be much simpler. Recall the alphabet we established for $\mathcal{L}_A$:

$$0, 1, 2, 3, 4, 5, 6, 7, 8, 9, +, \times, =, \leq, (,), x, ', \wedge, \vee, \neg, \Rightarrow, \Leftrightarrow, \exists, \forall, \&.$$

In our number scheme, each of these 26 symbols will have a unique *code number* (its Gödel number) that identifies it. In the coding presented here, each code number will consist of exactly *two digits*. The code numbers for the 26 symbols (in the order shown above) are just the *successive* two-digit numbers from 10 to 35. For example, the code number for each digit 0 through 9 is just that digit *prefixed* by the digit 1. So, these codes are:

$$10, 11, 12, 13, 14, 15, 16, 17, 18, 19.$$

Then, the code numbers for $+$ and $\times$ are 20 and 21; and the final code number, for symbol &, is 35.

With these code numbers, we can encode, as a *single* number, any expression E written in the alphabet of $\mathcal{L}_A$, where each symbol in E is simply replaced by its two-digit code number. This is our *Gödel number* of E, denoted $g(E)$. For example, the sentence "$\exists x(x + 10 = 62)$" would be (ignoring spaces) encoded

[11] "Gödel basically had to teach programming to formal systems that were about arithmetic. ... This is where all the tough technical work had to be done. The incompleteness theorems themselves were mere victory laps at the end." [81].

as a single number:

$$33262426201110221612 25$$

which is just the *concatenation* of the following code numbers for the individual symbols of the sentence:

$$33\ 26\ 24\ 26\ 20\ 11\ 10\ 22\ 16\ 12\ 25$$

Review question: Using the Gödel numbering scheme described above, what is the Gödel number of the predicate $x \leq 17$? What is the Gödel number of the predicate $x = 2 \times x$?

How many numbers are in the set of numbers expressed by the first predicate? What is the set of numbers expressed by the second predicate?

Encoding a derivation We will also need to encode an entire *derivation* (starting from axioms and running to a concluding expression) as a single Gödel number. For that encoding, we need the *end-of-expression* symbol, "&", to mark the end of each expression and separate it from the next one. Without such a symbol, two expressions might run together making it impossible to unambiguously recapture the derivation from its Gödel number.

Decoding Gödel numbers Since each symbol in the alphabet is encoded with two digits, the number of digits in the Gödel number for an expression will always be exactly *twice* the number of symbols in the expression (ignoring spaces). More important, given a Gödel number that encodes an expression, we can precisely *decode* it (left to right), reading two symbols at a time, unambiguously re-creating the expression that the Gödel number came from.[12] And note, the encoding and decoding can be easily done mechanically, by a machine or a computer program.

Review question: Decode the Gödel numbers

$$342627241110102326212426272011 2525$$

and

$$33263426272411101023262124262720112525.$$

[12] You might question whether the lack of encoding for spaces could cause some ambiguity. It does not, because all spaces can be removed from any expression in $\mathcal{L}_A$ without introducing any ambiguity. Or if you like, you can expand the alphabet to include a symbol for a space.

Determine, for both decoded expressions, whether it is a predicate, or a sentence, or neither? If it is a predicate, what set of numbers does it express? If it is a sentence, is it true or false in the normal interpretation of integer arithmetic?

9.3.2 Second Key Tool: The Diagonal Function $d()$

We are going to define a really critical, single-variable function, $d()$, called a *diagonal function*. I will explain the definition of $d()$ in three steps, but before that, I have to admit that I, and many others, find it hard to internalize this function. Still, by slowly defining $d()$ in a few steps, and drawing pictures, I can master the definition to the point that I can follow reasoning about it. So, don't be discouraged if you have to go over the definition of $d()$ a few times, and if you also never get any deep intuition about it – I haven't, but Gödel did! Pictures help.

Defining the diagonal function in three EZ steps (or payments)

First step Consider a predicate $E(x)$ in $\mathcal{L}_A$. Based on the *literal, explicit* symbols used in $E(x)$, the expression $E(x)$ has a Gödel number, say $\hat{n}$. As a *reminder* of the relationship of $E(x)$ to its Gödel number, we can also (optionally) use the notation $E_{\hat{n}}$ to refer to the predicate E. The notation $E_{\hat{n}}(x)$ is useful because it *reminds* us that $E(x)$ is a predicate whose Gödel number is $\hat{n}$.

As an example, suppose the predicate $E(x)$ is $x = 2 + 4$, which has Gödel number 2622122014. Then, $\hat{n}$ is 2622122014, and we can use the notation $E_{2622122014}(x)$ to refer to the predicate $x = 2 + 4$. Predicate $E_{2622122014}(x)$ expresses the set $\{6\}$, consisting of just a single number.

Second step Now replace x in $E_{\hat{n}}(x)$ with the number $\hat{n}$, that is, the Gödel number of the predicate denoted $E_{\hat{n}}(x)$. This turns the *predicate* $E_{\hat{n}}(x)$ into the *sentence* denoted $E_{\hat{n}}(\hat{n})$. It is a sentence because it now has no free variables. Further, the sentence is true if and only if it is satisfied by *its own* Gödel number, $\hat{n}$.

Continuing with our example, $E_{\hat{n}}(\hat{n})$ is the sentence denoted $E_{2622122014}(2622122014)$. In full detail, that sentence is

$$2622122014 = 2 + 4,$$

which is false (in our normal interpretation of integer arithmetic), but it is a sentence.

Third step Based again on the literal symbols used in the sentence denoted $E_{\hat{n}}(\hat{n})$, compute its Gödel number. That is, compute $g(E_{\hat{n}}(\hat{n}))$. We call the result $d(\hat{n})$, and call $d(n)$ a *diagonal* function.

Note that the function $d(n)$ is only defined when n has a value, $\hat{n}$, that is a Gödel number of some expression in $\mathcal{L}_A$.

> **Review question:** Compute $d(2622122014)$. That is, compute the Gödel number of the sentence $2622122014 = 2 + 4$.

Hopefully (if we both did it correctly) you found that

$$d(2622122014) = 121612121112121011422122014.$$

9.3.2.1 The Diagonal Function in a Nutshell

The function $d(n)$ may not be simple to state, and we have not yet seen why it is important, but it doesn't have to be completely opaque. In a nutshell, a value $d(\hat{n})$ is found as follows:

> Given a predicate $E(x)$, compute its Gödel number, $\hat{n}$; replace x with $\hat{n}$, creating the sentence $E_{\hat{n}}(\hat{n})$; compute its Gödel number, and call the result $d(\hat{n})$.

A visual overview of $d(n)$ To further cement the definition of $d(n)$, see Figure 9.1.

> **Review question** Is the following correct: $d(\hat{n})$, when defined, is the Gödel number of the sentence created by assigning the Gödel number, $\hat{n}$, of a predicate, to the variable in that predicate.
>
> Why might $d(\hat{n})$ not be defined?

> **Review question** Compute $d(34262724111010232621242627201 12525)$. Hint: Refer back to a previous Review question.

$$E(x) \rightarrow g(E(x)) = \hat{n} \rightarrow E(x) \equiv E_{\hat{n}}(x)$$
$$\rightarrow E_{\hat{n}}(\hat{n}) \rightarrow g(E_{\hat{n}}(\hat{n})) = d(\hat{n})$$

Figure 9.1 Function $d(n)$: The progression from the predicate $E(x)$ to the number $d(\hat{n})$ in the definition of function $d(n)$. The symbol "$\rightarrow$" here should be interpreted as meaning "leads to", and the symbol "$\equiv$" means "is equivalent to".

Review question Can you see why function $d(n)$ is called a diagonal function?

9.3.3 One Final Definition Before the Main Acts

For a set of Gödel numbers B $d^{-1}(B)$ is the set of all Gödel numbers n such that $d(n)$ is in B. So, any particular number $\hat{n}$ is in $d^{-1}(B)$ if and only if $d(\hat{n})$ is in B. The mapping d^{-1} is called the *inverse* of function $d()$.

Continuing with the above example, if the number 12161212111212 10111422122014 is in set A, then the number 2622122014 is in $d^{-1}(B)$, since

$$d(2622122014) = 1216121211121210111422122014.$$

See Figure 9.2.

All these definitions are hard to absorb – especially having little intuition for what they really mean. Figures 9.1 and 9.2 should help, but drawing your own figures to explain the definitions is what should help the most.

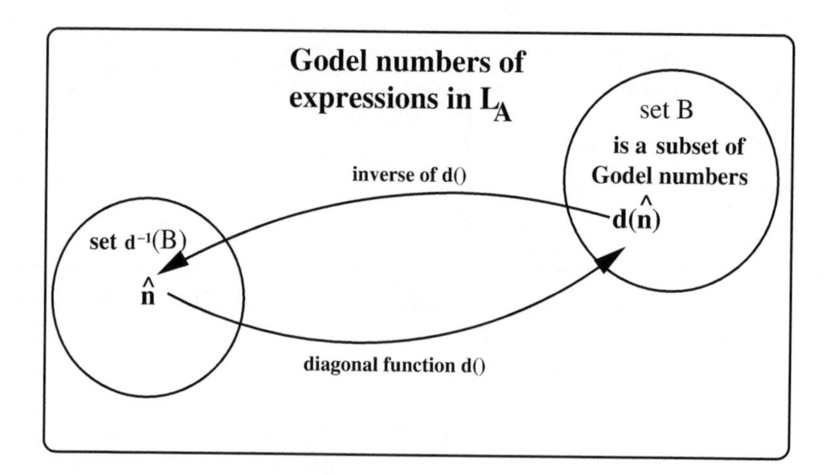

Figure 9.2 The relationship of set $d^{-1}(B)$ to set B. The enclosing box represents the set of all Gödel numbers of expressions in $\mathcal{L}_A$. Set B is a subset of Gödel numbers, and d^{-1} is a subset of Gödel numbers that are mapped to B by the diagonal function $d()$. The Gödel number 1216121211121210111422122014 is one element of set A, and 2622122014 is one element of set $d^{-1}(B)$.

9.4 A Diagonal Lemma

Now that you have mastered all these definitions (you have, haven't you?), we can prove a central Lemma about the diagonal function $d()$ and expressibility in a formal system Π containing language $\mathcal{L}_A$. We will then use the Lemma to prove an abstract version of Gödel's First incompleteness theorem.

Recall that a set of numbers, B is *expressible* in $\mathcal{L}_A$ if there is a predicate $E(x)$ in $\mathcal{L}_A$, such that any sentence $E(\hat{x})$ is true if and only if $\hat{x}$ is in B.

Lemma 9.4.1 *(A Diagonal Lemma) For a set of numbers, B suppose the set $d^{-1}(B)$ is expressible in $\mathcal{L}_A$. Then there is a predicate $E_{\hat{n}}(x)$ in $\mathcal{L}_A$ such that the sentence $E_{\hat{n}}(\hat{n})$ is* true *in $\mathcal{L}_A$ if and only if its Gödel number, $d(\hat{n})$, is in set B.*

The diagonal lemma is a mouthful. It might help to restate it:[13]

> If set $d^{-1}(B)$ is expressible in $\mathcal{L}_A$, then there is a *specific* predicate $E_{\hat{n}}(x)$ in $\mathcal{L}_A$ such that
>
> $$\text{Sentence } E_{\hat{n}}(\hat{n}) \text{ is true in } \mathcal{L}_A \iff d(\hat{n}) \in B.$$

Gödel sentences Note that the predicate $E_{\hat{n}}(x)$ defines a *set* of numbers, but $E_{\hat{n}}(\hat{n})$ is *one* specific sentence. That sentence is called a *Gödel sentence* for set B. A general definition of a Gödel sentence is:

> A Gödel sentence for a set of numbers, B, is a sentence that is true if and only if its Gödel number is in B.[14]

Gödel sentences are the critical elements (often paradoxical sounding and self-referential) in many incompleteness proofs, and in the proof of Gödel's Second incompleteness theorem.

> Almost a whole book could be written on various clever known methods for constructing Gödel sentences. [100]

Review question Does this general definition of a Gödel sentence agree with what was stated above about the sentence $E_{\hat{n}}(\hat{n})$? Explain in detail.

[13] Remember that the symbol " $\in$" means "is contained in".

[14] Quoting Smullyan [100]: "Informally, a Gödel sentence for B can be thought of as a sentence asserting that its own Gödel number lies in B."

Now we can give another statement of the diagonal lemma:

If set $d^{-1}(B)$ is expressible in $\mathcal{L}_A$, then there is a Gödel sentence for set B.[15]

9.4.1 Now We Prove the Diagonal Lemma

The proof of the diagonal lemma is really just a matter of applying and juggling definitions. So let's juggle!

Proof Assume that the set $d^{-1}(B)$ is expressible in $\mathcal{L}_A$. Then by the definition of *expressibility*, there is a predicate, call it $E(x)$, such that for any fixed value $\hat{x}$:

$$\text{The sentence } E(\hat{x}) \text{ is true in } \mathcal{L}_A \text{ if and only if } \hat{x} \text{ is in } d^{-1}(B). \tag{9.1}$$

Let $\hat{n}$ be the Gödel number of the predicate $E(x)$, which we can now call $E_{\hat{n}}(x)$. Then, replacing the variable x in $E_{\hat{n}}(x)$ with the constant number $\hat{n}$, statement (9.1) specializes to:

$$\text{Sentence } E_{\hat{n}}(\hat{n}) \text{ is true in } \mathcal{L}_A \text{ if and only if } \hat{n} \text{ is in set } d^{-1}(B). \tag{9.2}$$

Next, recall from the definition of set $d^{-1}(B)$, that

$$\hat{n} \in d^{-1}(B) \text{ if and only if } d(\hat{n}) \in B. \tag{9.3}$$

Putting (9.2) together with (9.3), we have:

$$\text{Sentence } E_{\hat{n}}(\hat{n}) \text{ is true in } \mathcal{L}_A \text{ if and only if } d(\hat{n}) \in B, \tag{9.4}$$

which is exactly what we wanted to prove. ■

The proof of the diagonal lemma is not hard to follow (line by line), but it is unlikely that you acquired much intuition about it. Still, knowing the lemma and having verified the proof, we can use it, as we will do next.

9.5 Now, an Actual Gödel Incompleteness Theorem

Now we state and prove an abstract version of Gödel's First incompleteness theorem (semantic version). Recall that we have assumed that a formal system, Π, contains the language of arithmetic $\mathcal{L}_A$. Recall also that Π is said to

[15] This statement of the diagonal lemma is a bit less informative than the prior two statements, because it doesn't specify the Gödel sentence. But, that is OK because the proof does specify it as $E_{\hat{n}}(\hat{n})$.

be *sound* if every sentence that can be derived in Π is *true* in the normal interpretation of integer arithmetic.

We let $\mathcal{U}$ denote the set of integers that are Gödel numbers for expressions in $\mathcal{L}_A$, and let $\mathcal{P} \subseteq \mathcal{U}$ denote the set of Gödel numbers of expressions in $\mathcal{L}_A$ that can be *derived* in Π. Then, $\widetilde{\mathcal{P}} \equiv \mathcal{U} - \mathcal{P}$ denotes the set of Gödel numbers of expressions in $\mathcal{L}_A$ that *cannot* be derived in Π. Of course, we don't know what integers are in sets $\mathcal{P}$ or $\widetilde{\mathcal{P}}$, but we can still reason about those sets.

Theorem 9.5.1 *(Gödel) Suppose a formal system Π contains the language of arithmetic $\mathcal{L}_A$, and that Π is sound. If the set of Gödel numbers, $d^{-1}(\widetilde{\mathcal{P}})$, is expressible in $\mathcal{L}_A$, then there is a* true *sentence that can be* written *in $\mathcal{L}_A$, but cannot be* derived *in Π, so Π is incomplete.*

9.5.1 And, the Proof

The proof is another exercise in juggling definitions, plus a little logic.

Proof Given the second sentence in the statement of the theorem, we assume throughout the proof that the set of Gödel numbers, $d^{-1}(\widetilde{\mathcal{P}})$, is expressible in $\mathcal{L}_A$. We call this the *expressability assumption*.[16] Then, the Diagonal Lemma (9.4.1) says that there is a predicate $E_{\hat{n}}$ in $\mathcal{L}_A$, such that:

The sentence $E_{\hat{n}}(\hat{n})$ in $\mathcal{L}_A$ is true if and only if its Gödel number is in $\widetilde{\mathcal{P}}$.

By the definition of $\widetilde{\mathcal{P}}$, a number is in $\widetilde{\mathcal{P}}$ if and only if it is the Gödel number of an expression that is *not derivable* in Π. Combining this with the statement above, establishes that the following statement is true:

$$\text{(Sentence } E_{\hat{n}}(\hat{n}) \text{ is true) } \textit{if and only if } (E_{\hat{n}}(\hat{n}) \text{ is } \textit{not} \text{ derivable in } \Pi). \quad (9.5)$$

Now recall that for two sentences, X and Y, the logical expression

$$X \text{ if and only if } Y$$

is *true* exactly when the values of X and Y *agree* (i.e., either both *true* or both *false*).

Applying this logical fact to the statement in (9.5), we replace sentence X with "(Sentence $E_{\hat{n}}(\hat{n})$ is true)", and replace sentence Y with "($E_{\hat{n}}(\hat{n})$ is *not* derivable in Π)". Then, since we have established that the statement in (9.5)

[16] We will not always remind the reader that we have made the expressability assumption, but anything we deduce to be *true* in the proof is based on that assumption.

is true, it follows that either both of these sentences are true, or both are false. This means that either:

a) Sentence $E_{\hat{n}}(\hat{n})$ is true and $E_{\hat{n}}(\hat{n})$ is *not* derivable in Π;
  or
b) Sentence $E_{\hat{n}}(\hat{n})$ is false and $E_{\hat{n}}(\hat{n})$ *is* derivable in Π.

But, by the assumption that Π is *sound*, and the definition of soundness, if $E_{\hat{n}}(\hat{n})$ is derivable in Π, it must be *true*. That rules out possibility *b*) in the two above possibilities. Hence, the only possibility is *a*), that the Gödel sentence $E_{\hat{n}}(\hat{n})$ is *true* but *cannot* be derived in Π. ■

Recapping: Given the expressability assumption, there is a true sentence (specifically $E_{\hat{n}}(\hat{n})$) in $\mathcal{L}_A$ that cannot be derived in Π, proving the abstract version of Gödel's First incompleteness theorem (semantic version).

9.5.2 A Gödel Sentence Was Used

Note that the sentence $E_{\hat{n}}(\hat{n})$ used in the proof of Theorem 9.5.1 is a *Gödel sentence* for set $\widetilde{\mathcal{P}}$. With that realization, we have the following:

Corollary 9.5.1 *When any sound formal system, Π, containing $\mathcal{L}_A$, has a Gödel sentence for $\widetilde{\mathcal{P}}$, it has a sentence that is true but not derivable in Π, so Π is incomplete.*

So, the key to proving incompleteness of a formal system Π is to find a Gödel sentence for $\widetilde{\mathcal{P}}$. This is an abstract *recipe* for proving Gödel incompleteness theorems in different formal systems.

Corollary 9.5.1 states a property of a Gödel sentence that goes beyond its previously stated definition. Misquoting from [100] (where I have replaced notation and terminology to be consistent with the exposition here):

> A Gödel sentence for $\widetilde{\mathcal{P}}$ is nothing more or less than a sentence which is true if and only if it is not provable [derivable] (in Π). And for any sound system Π, a Gödel sentence for $\widetilde{\mathcal{P}}$ is true but not derivable in Π. [Such a sentence can be thought of as asserting its own non-provability in Π.]

9.5.2.1 The Heart of It

Theorem 9.5.1 and Corollary 9.5.1 really get to the heart of Gödel's original approach to proving incompleteness and show the utility of Gödel sentences in establishing incompleteness. So, let's think a bit more about Gödel sentences in general. Recall the previously stated definition:

A Gödel sentence for a set of numbers, B is a sentence that is true if and only if its Gödel number is in B.

Essentially, what we did in the proof of Gödel's abstract incompleteness theorem (Theorem 9.5.1) was to use the fact (from the diagonal lemma) that: if $d^{-1}(B)$ is expressible in $\mathcal{L}_A$, then $\mathcal{L}_A$ has a Gödel sentence for B. Then, we specialized set A (in the definition of a Gödel sentence) to be the set $\widetilde{\mathcal{P}}$, giving:

If $d^{-1}(\widetilde{\mathcal{P}})$ is expressible in $\mathcal{L}_A$, then there is a Gödel sentence for the set $\widetilde{\mathcal{P}}$. That is, a sentence $E_{\hat{n}}(\hat{n})$ that is *true* in $\mathcal{L}_A$, if and only if its Gödel number, $d(\hat{n})$, is in $\widetilde{\mathcal{P}}$.

Equivalently, using the definition of $d()$:

If $d^{-1}(\widetilde{\mathcal{P}})$ is expressible in $\mathcal{L}_A$, then the sentence $E_{\hat{n}}(\hat{n})$ is *true* if and only if $E_{\hat{n}}(\hat{n})$ is *not derivable* in Π.

Then, since Π was assumed to be sound, we concluded:

If $d^{-1}(\widetilde{\mathcal{P}})$ is *expressible* in $\mathcal{L}_A$, then the sentence $E_{\hat{n}}(\hat{n})$ is *true*, but *not derivable* in Π.

For real this time We have proved an honest-to-goodness Gödel-stated theorem establishing conditions under which a formal system containing $\mathcal{L}_A$ will be incomplete. This is fine mathematics with great generality, but it doesn't establish the incompleteness of any *specific* formal system.

Getting specific In order to *use* Theorem 9.5.1 or Corollary 9.5.1 to establish the incompleteness of a particular formal system Π containing $\mathcal{L}_A$, we need to show that $d^{-1}(\widetilde{\mathcal{P}})$ is actually expressible in $\mathcal{L}_A$. We discuss that next.

9.5.3 Expressing $d^{-1}(\widetilde{\mathcal{P}})$: Another Three-Step Plan

We show how to express $d^{-1}(\widetilde{\mathcal{P}})$ in $\mathcal{L}_A$ by using the following three steps.

The first step is to show that set $\mathcal{P}$ is expressible in $\mathcal{L}_A$. The second step is to show that *if* $\mathcal{P}$ is expressible in $\mathcal{L}_A$, *then* $\widetilde{\mathcal{P}}$ is also expressible in $\mathcal{L}_A$. The third step is to show that *if* $\widetilde{\mathcal{P}}$ is expressible in $\mathcal{L}_A$, then so is $d^{-1}(\widetilde{\mathcal{P}})$. Taken together, these three steps show that $d^{-1}(\widetilde{\mathcal{P}})$ is expressible in $\mathcal{L}_A$.

These three steps can be accomplished by *directly* reasoning about $\mathcal{L}_A$ and Π, but that approach is extremely hard although the second step is trivial. Working directly in $\mathcal{L}_A$ involves a lot of grunge and forms a large part of what Gödel had to invent. So, I will not follow that approach. I will punt – but it is a *principled* punt, which will also introduce something very valuable.

Review question Explain why the second of the three above steps is trivial. That is, if $\mathcal{P}$ is expressible in $\mathcal{L}_A$, why must $\widetilde{\mathcal{P}}$ also be expressible?

9.5.3.1 Computation and Logic

After Gödel's original 1931 paper, when people (think Turing and others) came to understand more about the relationship of *computation* to logic, a powerful tool was elaborated that lets one establish that a set can be *expressed* in $\mathcal{L}_A$ by thinking in terms of computer-like *programming*, instead of thinking directly about $\mathcal{L}_A$. The tool is called *primitive recursion*, which is a *restricted*, yet powerful, form of computer programming.

Primitive recursion Informally, a *function* is called *primitive recursive* if its computation can be described (or prescribed) in a computer programming language where every loop (such as a FOR loop) iterates a number of times that is determined *before* the loop is executed. The number of times it iterates does not have to be specified at the time the program is *written*, but must be known before the loop is *executed*.

For example, the FOR loop with the header: "FOR (*i from* 1 *to* 5)" will iterate exactly five times and is fine. Similarly, the FOR loop with header: "FOR (*i from* 1 *to j*)" is fine as long as the value of variable j is determined before entry into the loop and will not change inside the loop. However, a loop with the header: "WHILE ($i < 10$)" is *not* fine *if* the value of i changes inside the loop in an unstructured way.[17] And, of course, a primitive recursive program can only use functions that are built into the programming language, *if* those functions themselves are primitive recursive.

Primitive recursion and Gödel numbers As an (optional) *illustration* of recursive programming, let's look at a primitive recursive program to compute the Gödel number, $g(x)$, of any *integer* x.

Input to the program will be a specific value, $\hat{x}$, assigned to variable x. With the Gödel numbering that we have been using in this chapter, the program has to insert the digit 1 before each digit in $\hat{x}$. We will use three functions that are known to be primitive recursive: $di(x)$ = the *number of digits* in a number x

[17] Quoting from [99]: "The crucial thing ... is that the required looping procedures are each iterated a fixed maximum number of times, set in advance [before] we first enter the loop (we can allow early exits). Contrast the open-ended searches involved in 'do until' (or 'do while') procedures, where we keep on looping around for as long as it takes, until (or while) some condition is satisfied."

PRIMITIVE RECURSIVE PROGRAM TO COMPUTE $(g(x))$

```
g = 0
m = 1
z = x̂
j = L(x̂)

for (i = 1, ..., j) do
   r = R(z, 10)
   z = W(z, 10)
   g = g + (r × m) + (m × 10)
   m = m × 100
endfor

g(x̂) = g
```

Figure 9.3 A primitive recursive program to compute the Gödel number, $g(x)$, given a positive integer $\hat{x}$.

(in detail: $di(x) = \lceil \log_{10}(x) \rceil$); $W(x, y)$, which is defined as the *whole* part of the division of x by y; and $R(x, y)$, defined as the *remainder* after dividing x by y. That is, $x = W(x, y) \times y + R(x, y)$. Then, the code shown in Figure 9.3 computes $g(x)$.

In the program, variable g accumulates the two-digit codes for each digit in z, *right to left*. At the end of the program, the value of g is $g(\hat{x})$. Variable z is initially given the value of $\hat{x}$, and j is assigned the number of digits in $\hat{x}$. In each of the j iterations of the FOR loop, the current z will be divided by 10. The remainder, $R(z, 10)$, is the least significant (rightmost) digit of z, and $W(z, 10)$ identifies the other digits of z. The variable m starts at 1 and is multiplied by 100 in each iteration; hence the value of m identifies the position in the growing $g(\hat{x})$, where the current value of r should be placed. Adding m puts the digit 1 one place to the left of where r is placed into $g(\hat{x})$.

The code is primitive recursive because the functions di, R, and W are known to be primitive recursive; and the FOR loop iterates $di(\hat{x})$ times, where $di(\hat{x})$ is a function of $\hat{x}$, which is known before the execution of the loop.

Primitive recursion and $\mathcal{L}_A$ The major theorem concerning primitive recursion and $\mathcal{L}_A$ is:

Theorem 9.5.2 *Any primitive recursive function can be* expressed *in* $\mathcal{L}_A$.

That is, if you can write a primitive recursive program to compute a function f, then you can craft a (two-variable) predicate $H(p, q)$ in $\mathcal{L}_A$ that expresses f.

In particular, for any values $\hat{p}$ and $\hat{q}$, the sentence "$H(\hat{p}, \hat{q})$" is true if and only if $f(\hat{p}) = \hat{q}$.

Proving this theorem involves a lot of hard work, detail, and grunge (it's grunge all the way down), and we will not even attempt it. But once the theorem has been established, you never again need to descend into the grunge – you can think at a higher level – in the realm of primitive recursive programming. Further, many common and useful functions have already been shown to be primitive recursive, and you can use any of those without having to descend into their grunge.

Back to $\mathcal{P}$ Recall that given $\mathcal{L}_A$ and Π, $\mathcal{P}$ is the set of all Gödel numbers (under our numbering scheme) of expressions in $\mathcal{L}_A$ that are *derivable* in Π.

Step one of our three-step plan is to show that set $\mathcal{P}$ can be expressed in $\mathcal{L}_A$. For that, we must construct a predicate *DERIVE()* such that for any Gödel number $\hat{x}$, $DERIVE(\hat{x})$ will be true if and only if $\hat{x}$ is the Gödel number of an expression in $\mathcal{P}$.

To begin, we *define* a two-variable predicate in $\mathcal{L}_A$, $\text{DER}(m, x)$ such that $\text{DER}(\hat{m}, \hat{x})$ is true if and only if $\hat{m}$ is the Gödel number of a *derivation* that ends with a sentence whose Gödel number is $\hat{x}$. Now, *defining* a predicate is not the same as *showing* that it can be written in $\mathcal{L}_A$ – but we will get to that shortly.

Assuming we can write the predicate $DER(m, x)$ in $\mathcal{L}_A$, predicate $DERIVE(x)$ can be written in $\mathcal{L}_A$ as:

$$\exists m(\text{DER}(m, x)),$$

so, for a fixed value $\hat{x}$ assigned to x,

$$\hat{x} \in \mathcal{P} \Leftrightarrow \text{DERIVE}(\hat{x}) \Leftrightarrow \exists m(\text{DER}(m, \hat{x})).$$

That is, $\hat{x}$ is the Gödel number of a derivable statement in Π if and only if there is a derivation (whose Gödel number is represented by the variable m) in Π of the statement with Gödel number $\hat{x}$ (Whew!).

So, it all comes down to whether we can express the predicate $DER(m, x)$ in $\mathcal{L}_A$. But how do we do that?

Primitive Recursion to the rescue, again To show that $DER(m, x)$ is expressible in $\mathcal{L}_A$, we exploit Theorem 9.5.2 and just show that $DER(m, x)$ is a primitive recursive function. That is, we need to construct a primitive recursive program that can take in any values $\hat{m}$ and $\hat{x}$ for m and x, respectively, and then determine whether or not $\hat{m}$ is the Gödel number of a derivation in Π of an expression in $\mathcal{L}_A$ whose Gödel number is $\hat{x}$.

That is a fairly straightforward task: First we *decode* $\hat{m}$ to completely spell out the sequence of symbols that $\hat{m}$ encodes. That is certainly doable by a primitive recursive program, just decoding pairs of digits from $\hat{m}$ left to right. This gives all of the symbols in the decoding of $\hat{m}$, but without any spaces. That is an easy problem to solve that we leave to the reader. The result is a string that might be a derivation in Π of an expression in $\mathcal{L}_A$ with Gödel number $\hat{x}$.

Now recall that in Chapter 6, we stated that derivations in any formal proof system Π must be *mechanically checkable* by a computer program that verifies that each statement in the derivation is either an axiom or a statement that follows from already established statements by the application of one of the rules of inference. Since the number of axioms is finite, and they are known to the program, checking if a statement is an axiom is clearly doable with primitive recursive programming. The number of rules of inference is also finite, and since we assumed in Chapter 6 that a derivation must specify the rule of inference used at any step, along with the statement(s) referred to in that application, checking if the step has correctly applied the rule of inference to existing statements in the derivation is easily done with primitive recursive programming. Finally, the verification must compute the Gödel number of the last statement in the derivation (which we have already seen can be done with primitive recursion programming) to check that it is $\hat{n}$.

In this way, the program can verify whether $\hat{m}$ is the Gödel number of a derivation in Π of an expression whose Gödel number is $\hat{n}$. If it is, then the program returns the value *true*, otherwise it returns the value *false*. Thinking of $DER(m, x)$ as a function that evaluates to *true* or *false* when given values $\hat{m}$ and $\hat{x}$, a primitive recursive program can determine the correct value for $DER(\hat{m}, \hat{n})$. Then, Theorem 9.5.2 implies that $DER(m, n)$ is a predicate expressible in $\mathcal{L}_A$. So, predicate *DERIVE()* is expressible in $\mathcal{L}_A$, and set $\mathcal{P}$ is expressible in $\mathcal{L}_A$. This completes the discussion of the First step of our three-step plan. The Second step is trivial, as we have already discussed.

Step three What remains is Step three: showing that if $\widetilde{\mathcal{P}}$ is expressible in $\mathcal{L}_A$, then so is $d^{-1}(\widetilde{\mathcal{P}})$. What that involves is creating a predicate in $\mathcal{L}_A$, let's call it $DINV(n)$, such that $DINV(\hat{n})$ is true if and only if $\hat{n}$ is a Gödel number such that $d(\hat{n})$ is in the set $\widetilde{\mathcal{P}}$.

> **Review question** Using primitive recursive programming and the predicate DERIVE(), and what we know about Step 2, complete the details for Step 3, showing how to express the predicate $DINV(n)$ in $\mathcal{L}_A$.

9.5.3.2 Recapping

We have given a complete, rigorous proof of Gödel's abstract (semantic) incompleteness theorem. The proof highlights the role of Gödel sentences and provides a roadmap for constructing them. Then, we gave a sketch of how the abstract incompleteness theorem can be applied when a specific formal system, Π, is specified, to show that Π is incomplete.

9.6 Gödel's Second Incompleteness Theorem

In a nutshell:

> If arithmetic is consistent, it cannot prove its own consistency. [100]

Before Gödel wrote a formal paper proving his First incompleteness theorem, he gave a talk about the theorem to a small group of mathematicians that included John von Neumann.[18] As the standard story goes, von Neumann was the only one in the audience who understood the talk, or seemed to care.

A little while after the talk, von Neumann wrote to Gödel to point out a consequence, now called Gödel's *Second Incompleteness* (or *Consistency*) Theorem, that many people believe is even more profound than Gödel's First incompleteness theorem.[19] Gödel wrote back to von Neumann that he had made the same discovery and intended to write up a full proof of it, but never did. However, he did state the result in the paper where he published his First incompleteness theorem.

> Gödel's proof of the second incompleteness theorem in his 1931 paper consisted mostly of hand-waving – in other words, sketching an argument without carrying it out in detail. [39]

In fact, a full formal proof of the Second theorem was only completed in 1939, by David Hilbert and Paul Bernays, and the formal proof is quite demanding.

Several other proofs of Gödel's Second incompleteness theorem have been developed, but none of these is really accessible to a broad audience the way that Gödel's First incompleteness theorem can be. So, currently only experts

[18] I was about to write "the famous mathematician John von Neumann", but at that time, von Neumann was only 27, and Gödel was only 25. Neither was famous yet, but both would become very famous, von Neumann perhaps more than Gödel. Recall that von Neumann was mentioned in Chapters 2 and 3.

[19] There is disagreement about which is more profound. People who know both theorems are divided into *firsters* and *seconders*. I am a firster, but there is no question that the Second theorem has deep, philosophical, and mathematical import.

have a full understanding of Gödel's Second theorem.[20] But, I still want to state Gödel's Second incompleteness theorem and give some intuition for it.

9.6.1 Informally

Informally, Gödel's Second theorem says that *if* a formal system Π is *rich-enough* and *consistent*, then there cannot be a derivation *inside* Π of any statement that establishes that Π is consistent, that is, that two contradictory statements can never be derived in Π.

The phrase "rich-enough" means that Π contains a language of arithmetic, axioms, and rules of inference that allow basic statements about arithmetic to be expressed and derived in Π. The formal systems that are widely used by mathematicians are all rich-enough in that sense.

Another (perhaps more informative) statement of Gödel's Second theorem is:

Theorem 9.6.1 *Any rich-enough formal system, Π, that can derive (inside Π) a statement implying that it is consistent is in fact inconsistent.*

Or as Scott Aaronson [1] puts it:

> ... the only mathematical theories pompous enough to [claim to] prove their own consistency are the ones that don't *have* any consistency to brag about![21]

9.6.2 It's an Inside Job

The most critical element of Gödel's Second incompleteness theorem is the idea of a formal system Π deriving a statement *totally inside* Π, that is, using only the language of arithmetic that is contained in Π, the axioms of Π, and the inference rules of Π. Franz puts this as follows:

> Central to the proof of the second incompleteness theorem is the notion of an ordinary mathematical proof being *formalizable* in a certain formal system. This means that for every step in the proof there is a corresponding series of applications of formal rules of inference in the system, so that the conclusion of the proof, when expressed in the language of the system, is also a theorem [a derived statement] of the system. [39]

[20] In writing this book, I read several discussions and proof sketches of Gödel's Second theorem, and tried to extract a rigorous proof that would be appropriate for the audience of this book. I failed. I don't think one exists. Yes, I have read [59]. It has a nice idea, but still relies on some very inaccessible heavy lifting – the Hilbert–Bernays derivability theorem (which they prove in 60 pages), that most hard-core mathematical logic books don't even try to prove.

[21] This statement is too sweeping, as we will see, but it does colorfully get across the essential point we want to make about Gödel's Second incompleteness theorem.

Satisfying the requirement that a derivation be formalized *totally inside* Π is the hardest part of the proof of Gödel's Second theorem, and the part that Gödel just "hand-waived" in his 1931 paper. Formalization wasn't completely shown until the work of Hilbert and Bernays, almost a decade after Gödel's paper. So, of course, we cannot fully explain the proof in this book. But, we can explain the high-level ideas in the proof and rigorously present two informative pieces of the proof. Those pieces are how to *express*, totally inside $\mathcal{L}_A$, the statement: "Π is consistent"; and how to state (but not derive), totally inside $\mathcal{L}_A$, the statement: "if Π is consistent then a particular statement, G_T, cannot be derived in Π." Later, we will say a bit about what statement G_T is.

Expressing consistency in $\mathcal{L}_A$ Recall that a formal system Π is said to be *inconsistent* if it can derive a statement and also derive its negation. So, it is clear how one can establish that Π is *in*consistent – just derive a statement and also derive its negation, in Π. But how can we ever express, and prove, that a formal system is *consistent*?[22] Certainly, in order to prove consistency in Π (that contains $\mathcal{L}_A$), you must be able to *express* consistency in the language $\mathcal{L}_A$. We will now show how to do that. The first thing we establish is:

Theorem 9.6.2 If *there is some statement* Q*, where both* Q *and* $\neg Q$ *can be derived in a formal system* Π*, then any statement* Y *that can be written in* Π *can be derived in* Π.

Proof First, observe that

$$X \Rightarrow (X \vee Y),$$

for any statements X and Y. In other words, if a statement X is true, and Y is *any* statement, then the statement "X OR Y" must also be true. This is due to the meaning of OR, and more formally, by the meaning of the logical symbol '$\vee$' in $\mathcal{L}_A$, as explained in Section 9.2.1.

Now, let's take X to be the statement

$$(Q \wedge \neg Q),$$

so applying the above, for any statement Y,

$$(Q \wedge \neg Q) \Rightarrow ((Q \wedge \neg Q) \vee Y).$$

But

$$(Q \wedge \neg Q)$$

[22] As the famous saying goes: "You can't prove a positive!"

evaluates to *False*, and so

$$((Q \wedge \neg Q) \vee Y)$$

is equivalent to Y. That is,

$$((Q \wedge \neg Q) \vee Y)$$

will be true if and only if Y is true. Combining the pieces, we have shown that

$$(Q \wedge \neg Q) \Rightarrow ((Q \wedge \neg Q) \vee Y) \Leftrightarrow Y,$$

so

$$(Q \wedge \neg Q) \Rightarrow Y$$

is a true statement. ■

Summarizing, an inconsistent formal system Π can derive *every* statement that can be formed in Π. All this seems very technical, but it has a really practical consequence:

Theorem 9.6.3 *If there is even one statement that can be formed but not derived in a formal system* Π *(that contains* $\mathcal{L}_A$*), then* Π *must be consistent.*

So, at a high level, it is *conceptually* easy to establish that such a formal system Π is *consistent*: Just prove that there is a statement that can be formed in $\mathcal{L}_A$, but *cannot* be derived in Π. And, if Π is *sound* (as we want it to be), no *false* statement should be derivable in Π. The obvious choice, then, for a statement in $\mathcal{L}_A$ that cannot be derived in Π is one that is false. A commonly used statement is "1 = 0", which looks pretty darn false to me. Summarizing:

Theorem 9.6.4 *A sound formal system* Π *containing* $\mathcal{L}_A$ *is consistent if and only if the sentence* "1 = 0" cannot *be derived in* Π.

9.6.3 Say It in $\mathcal{L}_A$

We want to make the assertion *totally inside* $\mathcal{L}_A$ of: "The sentence '1 = 0' cannot be derived in Π." We have already discussed all the needed elements for this. First, the Gödel number for "1 = 0" (in the numbering scheme discussed in Section 9.3) is 112210, the concatenation of the individual codes 11, 22, and 10, for the symbols "1", "=", and "0". Then, using the predicate *DERIVE* discussed earlier, "*DERIVE*(112210)" asserts that the statement "1 = 0" *can* be derived in Π. So, of course, we need to change it to:

¬ *DERIVE*(112210),

which is the desired statement in $\mathcal{L}_A$ that implies that Π is consistent.

9.6.4 Proof of the Second Theorem: Gödel Sketcher Is Back[23]

Although we cannot present a complete, rigorous proof of Gödel's Second Theorem, we can *sketch* it's *high-level* outline and non-formal ideas.

The basic idea is straightforward, the devil is in the details of a full proof [99].

The canonical Gödel sentence In Section 9.4 we defined Gödel sentences and used one in the proof of the Diagonal Lemma (Lemma 9.4.1). Gödel, in his original paper, defined a language of arithmetic, which we will pretend is our $\mathcal{L}_A$ (after all, this is just a sketch). He then described a particular Gödel sentence that is often denoted G_T, and is considered as *The Canonical* Gödel sentence.

We assume throughout that the formal system Π is "rich-enough". Then, the following seven points sketch the high-level ideas in the proof of Gödel's Second theorem:

1. Sentence G_T can be informally *interpreted* as saying: Sentence G_T is not derivable in Π.
2. Gödel's *First* Incompleteness Theorem (syntactic version) essentially says: If a formal system Π is consistent, then the sentence G_T is *not* derivable in Π.
3. Points 1 and 2 then imply that statement G_T is *true* in Π, if Π is consistent.

> **Review question** Fully explain why points 1 and 2 together imply that statement G_T is true in Π.

4. Close inspection of the Gödel's original *proof* of his *First* Incompleteness Theorem (syntactic version) shows that all of the statements in it can be written in the language $\mathcal{L}_A$, and that the inferences in the proof only use the inference rules given in Π.
5. Points 3 and 4 imply that the statement "If Π is consistent, then sentence G_T is true" can be derived *inside* Π.
6. Hence, *if* the statement "Π is consistent" can *also* be derived *inside* Π, then combined with Point 5, we would have a derivation, inside Π, of the sentence G_T.

[23] Bad pun! Apologies to Douglas Hofstadter.
I really tried to find a way to title this section "Gödel Etcher is Bach", but that was too much of a stretch.

7. But, Gödel's *First* Incompleteness theorem says G_T cannot be derived in Π, provided that Π is consistent (see Point 2). That, combined with Point 6 leads to the conclusion:

   > If Π is consistent, then the statement "Π is consistent" *cannot* be derived inside Π,

   which is Gödel's *Second* incompleteness Theorem.

   An immediate, and very significant Corollary is:

Corollary 9.6.1 *If the statement "Π is consistent"* can *be derived in Π, then* Π must *actually be* inconsistent.

As pointed out in [81], this Corollary establishes a truth that is analogous to: The act of saying that you are a *very stable genius* is absolute proof that you are not one.

I want to stress this deep, subtle point. Gödel's Second Theorem does not just imply that a "rich-enough", consistent proof system cannot prove (i.e., establish the truth) that it is consistent – it cannot even *derive* a statement implying that it is consistent. It is as though a consistent Π is *tongue tied.* Another analogy is that only vain people can say that they are modest – it would be a logical contradiction for a truly modest person to even utter those words.

9.6.5 Another Way of Explaining the Proof

Recall the *First incompleteness theorem (syntactic version):* If a "rich-enough" formal system Π (which contains $\mathcal{L}_A$) is consistent, then statement G_T is true, but cannot be derived in Π.

In proving the First incompleteness theorem, Gödel was allowed (as were we) to use any language and any mathematically acceptable logic. He was not limited to using a language of arithmetic or limited to using inference rules in the formal system to which the First theorem applied. He could use any mathematically accepted language and logic.

However, von Neumann pointed out, and Gödel himself had realized, that the entire argument he used in his original proof of the First incompleteness theorem could actually be written in the language of arithmetic, $\mathcal{L}_A$, and never use logical rules outside of certain commonly accepted formal systems (in particular, a variant of the formal system used by Russell and Whitehead in *Principia Mathematica*) which were "rich-enough". Doing this is very tedious, and we will not attempt it. Gödel didn't either, but he saw (and others accepted) that it could be done:

> ... we need to verify that the proof of the implication "if Π is consistent then G_T is unprovable in Π" is indeed formalizable in Π. In his paper, Gödel only presented this as a plausible claim, ... [39]

The punchline Suppose next that instead of *assuming* consistency (as was done in the statement of the First incompleteness theorem), the consistency of Π could be formalized and derived *inside* Π. That derivation would be a series of statements in $\mathcal{L}_A$ that only use the logical inference rules and axioms of Π. Then, we could take that derivation and follow it with the derivation, inside Π, of "if Π is consistent, then G_T is true". Those two statements together would be a proof, entirely inside Π, that G_T is *true*. But that means that statement G_T *would* be derived *inside* Π, *contradicting* Gödel's First incompleteness theorem, that is, that G_T *cannot* be derived in Π.

The only way to resolve this contradiction is to conclude that the original assumption, that the consistency of Π can be derived inside Π, is not correct. Hence, this proves Gödel's *Second* incompleteness theorem.

9.6.6 What Does "Rich-Enough" Really Mean?

We have used the phrase "rich-enough" many times, and said that it means Π contains a language of arithmetic, axioms, and rules of inference that allow basic statements about arithmetic to be expressed and derived in Π. But how much basic arithmetic is necessary?

We have been vague about this, and we will continue to be vague. However, there are two points to note. First, quoting from [99]:

> ... I'll not pause here to spell out just how much arithmetic that is, though ... it is *stunningly little.* (italics added)

Second, although the amount of arithmetic needed is "stunningly little" for both the First and Second theorems, the amount of basic arithmetic that must be expressed and derived in a formal system is greater for the proof of Gödel's Second theorem than for the First theorem.[24] So, there are formal systems that

[24] We will not detail these differences but only introduce a word, *induction*, in case you want to delve deeper into this issue. For Gödel's Second theorem, a formal system must be able to express and derive statements that concern *mathematical induction*. That is not required for Gödel's First incompleteness theorem.

are rich-enough for Gödel's First theorem to apply but not rich-enough for the Second theorem to apply. All of the formal systems that are commonly used by mathematicians are rich enough for both theorems to apply.

9.7 Misconceptions about Gödel's Theorems

There are many misconceptions and misstatements about Gödel's theorems, both the First and Second.

> Many references to the incompleteness theorem outside the field of formal logic are rather obviously nonsensical and appear to be based on gross misunderstandings or some process of free association. [39]

Many of these references state that the theorems have philosophical, societal, or religious implications well beyond their technical meanings. One such misconception (or actually, abuse) was quoted in Chapter 1. Two books [16, 39] discuss many more such abuses, and I don't want to discuss *non-mathematical* misconceptions here. But, I do want to discuss several misconceptions about the *mathematical* content of Gödel's theorems and proofs.

9.7.1 Misconceptions about the First Theorem

The following is an example of a common misstatement about Gödel's First incompleteness theorem:

> ... all mathematical systems contain statements which are true, yet cannot be proved within the system.[25]

The error in this statement is that Gödel's First theorem does *not* apply to *all* mathematical systems. In order for Gödel's theorem to imply that a formal system contains a statement that is true, but unprovable in the system, the system must be "rich-enough", as discussed earlier.[26]

A related, but more technical misstatement is:

> There are *no* formal systems that are both *consistent* and *complete*.

[25] This statement may remind you of the correct popular statement given at the start of Chapter 6, but that statement contains the catch-all qualifying phrase: "for systems to which the theorem applies".

[26] This misstatement is particularly ironic, because it appears on the back cover of [16], while the actual content of that book shows the ways that this common statement is wrong. You may wonder how this can happen – in most books, the back cover material is not written by the author, or even reviewed by the author before publication. I also had misstatements on the back covers of two of my previous books, which fortunately were relatively harmless.

This is wrong, as reflected in the following quote:.

> [Gödel's first] incompleteness theorem does not imply that *every* consistent formal system is incomplete. On the contrary, there are many complete and consistent formal systems. [39] (italics are original)

That is, there are many formal systems where for every statement that can be formed in the system, either that statement or its negation can be derived (and so the system is complete), and where it is not possible for a statement and its negation to both be derived (so the system is consistent). In fact, Alfred Tarski showed that for ordinary high-school algebra, and for Euclidean geometry (the standard geometry that is taught in school), there is a formal system that is consistent and complete:

> Tarski brought a ray of sunshine, by discovering a theory weak enough that Gödel's theorem does not apply, yet strong enough to formalize high-school algebra and Euclidean geometry.
>
> ... so incompleteness applies to number theory, but not to algebra and geometry, at least, ordinary Euclidean geometry and high-school algebra. [12]

An even more extreme, but common, misstatement is:

> There are true statements in arithmetic that cannot be proved in any formal system.

As discussed earlier, in Section 6.6, this misstatement reverses the direction of implication in a correct summary of Gödel's First incompleteness theorem (semantic version), in addition to not saying that the formal system must be "rich-enough".

The misconception on steroids:

> There are truths in mathematics that cannot be proven.

Definitely not what Gödel established. Hopefully, by now, no further discussion of this is needed.

Gödel's *Incompletability* Theorem Another misconception about Gödel's First incompleteness theorem is that the incompleteness of a formal system Π can be cured by adding more axioms to Π. The following quote from [55] hints at this possibility:

> ... when a proposition is undecidable in such a [formal] system, that means only that its axioms do not provide enough information to decide it. But new axioms, external to the original set, might supply the missing information and make the proposition decidable after all.

Actually, the above statement is correct (in fact, we could just make any particular proposition an axiom in the modified Π), but the larger suggestion is wrong. After adding a new proposition to the axioms, there will still be other propositions that are undecidable in the system, meaning that the incompleteness of the system will not be cured: once incomplete – always incomplete.

The reason is that if a system is "rich-enough", so that Gödel's First incompleteness theorem applies, then after adding more axioms, the system will still be "rich-enough" and so Gödel's incompleteness theorem will continue to apply. With that reality, one might call the Gödel's theorem an *Incompletability* theorem rather than an incompleteness theorem.[27]

9.7.2 Misconceptions about the Second Theorem

The main misstatement The following is a typical and common type of misstatement:

> ... no mathematical system can provide a proof of its own consistency. [55]

The error is that the statement is too sweeping. Gödel's Second theorem applies to *certain* formal systems, ones that are "rich-enough". Generally, formal systems that are of real interest and importance to working mathematicians *are* rich-enough, so Gödel's Second theorem applies to them. But, it is not true that Gödel's Second theorem applies to *all* mathematical systems. In fact, there are formal systems that *can* prove their own consistency, although those systems tend to be fairly weak, that is, limited in what can be derived in them.

An extreme misconception Quoting Smullyan [100]:

> We have seen such irresponsible statements as "By Gödel's second theorem we can never know whether or not arithmetic is consistent." Rubbish!

The error is that Gödel's Second theorem concerns what can be established *inside* a formal system. The theorem does not limit what can be established by methods and logics *outside* of the system, and so the Theorem does not "constitute the slightest rational grounds for doubting the consistency of P.A." [100][28]

[27] It is somewhat counter intuitive that to avoid incompleteness, one has to make a system *weaker*, that is, able to derive *less*. Incompleteness creeps in when systems become stronger.

[28] *P.A.* (Peano Arithmetic) is a widely used formal system axiomatizing arithmetic. It is generally believed to be strong enough to prove all the standard results in number theory, for example, all the known results concerning prime numbers.

A relatively harmless, but misleading, misconception Some lay-audience expositions say that Gödel's Second theorem follows *easily* from the First theorem. For example: "We've learned that if a set of axioms is consistent, then it is incomplete. That's Gödel's First incompleteness theorem. The second – that no set of axioms can prove its own consistency – easily follows" [108].

That claim is false. It makes the error of being too sweeping, but the main error is the claim that Gödel's Second theorem *follows* (easily or not) from Gödel's First theorem. If it were true, surely Gödel would have seen it and would have said so. But, "Gödel himself didn't prove [it] ... The hard work was first done by David Hilbert and Paul Bernays in .... (1939)" [98].

Gödel's Second incompleteness theorem does not follow from the *statement* of Gödel's First incompleteness theorem.[29] The following claim is more defensible:

> The second incompleteness theorem follows from Gödel's *original proof* for the first incompleteness theorem. [59] (italics added)

A more informative claim is that Gödel's proof (augmented by Hilbert and Bernays) of the Second incompleteness theorem exploits and builds on certain elements of Gödel's original *proof* of the First theorem.

I said that this misconception is relatively harmless, but I did waste more time than I want to admit to, trying to verify that the Second theorem is a *corollary* of the First theorem – it isn't.[30] I suspect others have wasted time as well.

9.7.2.1 A Perplexing Technical Misconception

I am truly fascinated by a technical misconception written by a prominent physicist and probabilist E. T. Jaynes (referred to as "the brilliant physicist E.T. Jaynes" in [47]), who stated that his views originated with two famous English scientists, Sir Ronald Fisher and Sir Harold Jeffreys.

We start with something that is true (and essentially known since ancient times). Recall from Theorem 9.6.2 that an inconsistent formal system Π can derive *every* statement that can be written in Π. Hence, if we don't know whether Π is consistent or not, what could we learn, *by itself*, from a derivation in Π of a statement saying that Π is consistent? *Nothing*! If Π is inconsistent, it can derive the same statement. It is like an accused person claiming they are innocent – both guilty and innocent people say that, so hearing that claim tells you nothing about whether the person really is guilty or innocent. Those

[29] When a result B follows from the *statement* of a result A, B is called a *Corollary* of A.

[30] Unfortunately, Gödel's Second theorem is called a corollary of Gödel's First theorem in [16]. As explained above, that is a misuse of the term "corollary".

words convey *no* information. Let's call this the *trivial fact*. For emphasis, here is how the *trivial-fact* is discussed in Wikipedia:

> It would actually provide no interesting information if a system F proved its consistency. This is because inconsistent theories [formal systems] prove everything, including their consistency. Thus a consistency proof of F in F would give us no clue as to whether F really is consistent; no doubts about the consistency of F would be resolved by such a consistency proof.[31]

Another (forceful) statement of the *trivial-fact* is written in [100] on page 109.

OK, but so what? The *trivial-fact* is true, but what does it have to do with Gödel's Second incompleteness theorem? Define T as the statement: "Π is consistent". Here is how Jaynes [55] states the *trivial-fact*:

> ... the fact that T can be deduced from the axioms [of a system] cannot prove that there is no contradiction in them, since, if there were a contradiction [so the system is inconsistent], T could certainly [also] be deduced from them!

Jaynes's statement of the *trivial-fact* is fine, but the statement continues:

> This is the essence of Gödel's theorem, ... As noted by Fisher (1956) [38], it shows the intuitive reason why Gödel's result is true. We do not suppose that any logician would accept Fisher's simple argument as a proof of the full Gödel theorem; yet for most of us it is more convincing than Gödel's long and complicated proof.

So Jaynes is claiming that the essence of Gödel's (Second) theorem is just the *trivial-fact*, and that this explanation originates with Sir Roland Fisher and Sir Harold Jeffreys. According to Jaynes, the *trivial-fact* is why Π can't prove its own consistency: a "proof" in Π that Π is consistent would give us *no* information about whether Π is really consistent or not, and so the consistency of Π cannot be proved in Π.[32]

Wrong, wrong, wrong! Let's recall the words at the start of this discussion:

> ... if we don't know whether Π is consistent or not, what could we learn, *by itself*, from a derivation in Π of a statement saying that Π is consistent? Nothing!

[31] Remember that logicians usually use the words "prove" and "proof" when we have used the words "derive" and "derivation". So, "prove" in this quote does not mean "establish the truth of". It means "derive". But, in the upcoming quote from Jaynes, "prove" means "prove", and "deduced" means "derived".

[32] Actually, this reasoning can be defended, *if* by "prove" one means "establish the truth of". But, in logic, the word "prove" just means "derive", unconnected to any claim of "truth".

The key phrase in this "obviously true" statement is "*by itself*". If you have no other tools or information to bring to bear on the question of consistency, the trivial-fact alone *does* invite the belief that a derivation in Π of a statement saying that Π is consistent would *not* resolve the question. But, that totally ignores Gödel's Second theorem, which not only provides another tool, it provides a nuclear-powered sledge hammer:

> If a rich-enough system Π can *derive* a statement implying "Π is consistent", then Π is absolutely *not* consistent.

Contrary to what Jaynes implies (i.e., that a "proof" in Π that Π is consistent would give us *no* information), such a proof (and here I mean a derivation), if it existed, would give us (because of Gödel's Second theorem) *every bit* of information about whether Π is consistent or not – it is *not*!

That deep, difficult, non-trivial result is why the Second theorem has a "long and complicated proof", and why (damn straight!) "no logician would accept Fisher's simple argument as a proof of the full Gödel theorem". The trivial-fact does not make Gödel's Second theorem trivial – as Jaynes though it did – it does the opposite. It highlights how enormously deep and non-trivial is the power of Gödel's pure reason. What could be a better proof (in the colloquial sense) of the power of mathematics and pure reason than to prove conclusively that a "trivially true" belief is completely wrong?[33]

This confusion of the *trivial-fact* for Gödel's deep and profound Second theorem is really interesting to me, because a) it is *fundamentally, deeply* wrong; and b) "brilliant" Distinguished Professor E. T. Jaynes, was a very prominent, accomplished, mathematically oriented scientist – not a journalist or a science writer or an amateur mathematician; he was quite capable of understanding what Gödel's Second theorem actually says, even if he didn't understand the proof.

[33] I found Jaynes's book containing his discussion of Gödel during a search for Gödel on Google. The assertion in his book that his understanding of Gödel's theorem comes from Fisher in no typo – the assertion is made several times. For example, in Jaynes's annotated bibliography, the entry for Fisher's 1956 book states: "He shows his old power of intuitive insight in his neat explanation of Gödel's theorem."

But, I have searched every page of Fisher's 1956 book, and the 1959 second edition, and find absolutely no mention of Gödel in Fisher's book. Further, I have read Jeffreys's 1957 comments on Gödel, and I find them to be correct – they do not say or suggest that the essence of Gödel's Second theorem is the trivial-fact. So, it seems to me that Jaynes not only misunderstood Gödel's second theorem, and misunderstood Jeffreys's comments about it, but Jaynes was also mistaken that Fisher wrote about Gödel in his 1956 book.

I have never seen this little bit of history discussed before. This "historical find", which may be of interest to historians or philosophers or sociologists of science, is my only academic contribution to any of the topics in this book.

Why so wrong? So, how could Jaynes have gotten it so wrong? I have tried to imagine what he must have thought of Gödel if he thought that Gödel's famous (second) incompleteness theorem was actually a trivial matter. Did he really dismissively think that Gödel and all those other logicians and philosophers, with their "stilted jargon ... linguistic tricks and ... meta-language gobbledygook" [55], were just pompous windbags? Based on his use of such dismissive language, maybe he did, and that gave him the confidence (or arrogance) to believe that he didn't need to consult with any logicians to check out his understanding of Gödel's theorems. I would really like to know.[34]

9.7.3 Misconceptions about the Causes of Incompleteness

Gödel's *original* proofs of his First and Second incompleteness theorems use three concepts that have gotten a great deal of attention in the popular press: *Gödel numbering based on prime powers, Explicit self-reference, and Twisty, paradox-based Gödel sentences.*

These three concepts are heavily emphasized in most (if not all) popular-math expositions of Gödel's incompleteness. Unfortunately, that focus gives an incorrect impression of the incompleteness phenomenon and its causes, often adding to the mystical status of Gödel's proofs.

9.7.3.1 Gödel Numbering Based on Prime Powers

As mentioned earlier, Gödel created a coding (a Gödel numbering) that is much more complex than that presented in Section 9.3.1. His Gödel numbering is based on non-trivial properties of *prime numbers*.[35]

[34] I have been highly critical of Jaynes's error and his dismissive attitude toward logicians and philosophers. Of course, I should not throw stones, because it is possible (even likely) that I have made errors in my understanding of some things in this book. But if so, it certainly is not because I think the discovers of the theorems and models were dolts.

[35] Very briefly, suppose that the alphabet of the language of arithmetic is the alphabet we have used, listed in Section 9.2, and suppose we code each symbol in the alphabet by its position in that ordered list. So, the digits 0 to 9 would be coded with the numbers 1 through 10, and the next symbol, '+' would be coded with the number 11, etc. Then, Gödel's original Gödel number of a statement of length n is formed by taking the first n prime numbers, starting from 2, raising each one to the power of the code of the associated symbol in the statement, and then taking the product of all of these numbers. For example, the statement "7 = 3 + 4" would have Gödel number

$$2^8 \times 3^{13} \times 5^4 \times 7^{11} \times 9^5,$$

whatever that comes to. Decoding is possible because of a basic theorem in number theory, the *unique factorization theorem* that says that for every integer, there is only one way to write that integer as the product of powers of prime numbers. So, given the Gödel number of a statement, that statement can be correctly re-created. In Gödel's original proofs, his Gödel numbers are also used in a way that requires the application of a more advanced theorem in number theory called the *Chinese Remainder Theorem.*

Now, Gödel's Gödel numbers become *HUGE*.[36] And with Gödel's numbering scheme, decoding a Gödel number to get back the originating statement, while always possible, is not an easy task. But those practical issues are irrelevant to Gödel's *theorems*, although they do add to their mystique, as does Gödel's use of the Chinese remainder theorem from number theory. Some people have conjectured that by requiring such huge numbers, Gödel's theorems might be irrelevant in physical systems (e.g., the brain) that can't represent those numbers.

As we mentioned in Section 9.3.1, Gödel used such a complex numbering scheme in order "program without programming", which we do not need to do, since we can do real programming. The only real requirement for a Gödel numbering scheme is that statements in $\mathcal{L}_A$ must be coded into integers in a way that they can be uniquely decoded, and both operations (coding and decoding) must be doable by primitive recursive programs [98]. Moreover, there are incompleteness results that do not use any kind of Gödel numbering.[37]

Gödel's original coding scheme is a *relic*, perhaps of historical interest, but certainly not essential to any incompleteness phenomena. More generally, Gödel's "arithmetization of syntax" is used to prove incompleteness of formal systems concerned with *arithmetic*, but is not always required for incompleteness in other contexts.

OK, so what? What I find very disappointing is that most general-audience expositions of Gödel's theorems, and mathematics and logic courses, *still* (more than 90 years after Gödel's major paper) emphasize the central role of Gödel's original numbering scheme – conveying the impression that this complex, and highly mathematical detail, like a "mystical, priestly incantation", is a necessary cause of incompleteness. But, it is not true:

> People always say, "the proof of the Incompleteness Theorem was a technical tour de force, it took 30 pages, it requires an elaborate construction involving prime numbers," etc. Unbelievably, 80 years after Gödel, that's still how the proof is presented in math classes! [1]

[36] Using Gödel's numbering scheme, but with a different ordering of the symbols in the alphabet than the ordering given in this chapter, the statement "$0 = 0$" has Gödel number 243,000,000 [77].

[37] The incompleteness results in Chapters 6, 7, and 8 do not use Gödel numbers, and the first presented incompleteness theorem in [98] appears almost 100 pages before Gödel numbers are even introduced.

9.7.3.2 Explicit Self-Reference

An example of *explicit self-reference* is the following statement that we will call *Statement Z*:

> Statement Z is not derivable in formal system Π.[38]

Gödel did use explicit self-reference even though it can be avoided. The abstract version of Gödel's First incompleteness theorem discussed in this chapter (which follows Gödel's actual writing) does involve explicit self-reference, in the definition of function *d*() and in the Diagonal Lemma. But, the versions of incompleteness that we established earlier in Chapters 6, 7, and 8 do *not* use explicit self-reference. The diagonalization used in Chapters 6 and 7 may be viewed as an *implicit* form of self-reference, but not an *explicit* form. The incompleteness proved in Chapter 8 does not use *any* form of self-reference. These versions of incompleteness show clearly that incompleteness does *not* depend on explicit self-reference,[39] although it is central in Gödel's *original* proofs. Commenting on an incompleteness proof presented in [39], Franzén writes:

> This version of the first incompleteness theorem based on computability theory shows that the use of (arithmetical formulations of) self-referential sentences is not essential when proving incompleteness. Indeed, there is no assumption in the proof that [Π] is capable of formalizing self-reference. [39]

Self-reference and the Mind Many popular science expositions and (inappropriate) exploitations of Gödel (particularly, attempts to connect Gödel to issues of the brain and mind) emphasize self-reference as the mystical, almost-magical, key element of Gödel's proofs. But, as the incompleteness proofs in Chapters 6, 7, and 8 show, explicit self-reference is *not* an essential ingredient of Gödel and Gödel-like incompleteness (although it is sometimes mathematically convenient to use it):

[38] Note that Statement Z explicitly refers to itself. Statement Z can be used to establish an incompleteness theorem as follows: If Z can be derived in Π, then what Z states is *false*, so Z isn't sound. Hence, if Z can be *written* in Π, and Π is sound, we can conclude that Z can't be *derived* in Π. But, if Z can't be derived in Π, then statement Z is *true*, so Π contains a statement that is true but not derivable in Π. Then, to fully prove this incompleteness, we need to show how statement Z can be written in Π. That is often where the hard work is.

I really hate this type of reasoning, using *explicit* self-reference, because it makes my head hurt, and I never really feel sure that we aren't just playing word games. That is why I have avoided explicit self-reference as much as possible in this book. But this kind of explicit self-reference is central to the mystique of Gödel's proof, particularly in popular science expositions.

[39] The first presented incompleteness theorem proved in [98] also does not use explicit self-reference.

> *...there are formally undecidable sentences which aren't self-referential.* That observation ought to scotch once and for all any lingering suspicion that the incompleteness phenomena are somehow inevitably tainted by self-referential paradox (italics are in the original quote). [98]

Self-reference mystery-mongering can be, and should be, avoided!

9.7.3.3 Twisty Gödel Sentences and Paradoxes

We earlier discussed Gödel's use of what are now called Gödel sentences and noted that the *The Canonical* Gödel sentence G_T is informally interpreted as saying:

> Sentence G_T cannot be derived in Π,

which is a self-referential statement. In most general-audience expositions of Gödel's theorems, the very twisty sentence G_T is motivated and compared to well-known paradoxes such as the self-referential Liar's paradox: "Everything I say is a lie".

> What Gödel managed to do was to construct a rigorous mathematical version of the liar paradox using only basic arithmetic. [29]

But other paradoxes that are not self-referential can also be used in deriving Gödel's incompleteness theorems. One example is

> **The Berry Paradox** If we limit the number of characters (from a finite alphabet) used in specifying an integer to some fixed number (say 72), then there can only be a finite number of integers specified. So, there must be a *smallest integer that cannot be specified using fewer than 73 characters*. But, that phrase specifies a particular integer, and only uses 72 characters (if I counted right). So, there is no such smallest integer!

This seems to lead to the conclusion that using only 72 characters does *not* limit the number of integers that can be specified, since there is no smallest integer that cannot be so specified. But, we already established that there are only a finite number of integers that can be specified using at most 72 characters. So, this is a paradox, but it is not self-referential,[40] and it has been used as the basis for proofs of Gödel's theorems [17, 93].

Given the above, and the emphasis in the popular science press on the role of paradoxes in Gödel's theorems, the next natural question is whether any paradoxes, or twisty Gödel sentences are necessary for incompleteness. In fact, we have already seen that the answer is "No": The incompleteness results in

[40] Although, the Wikipedia article on Berry's paradox says it is. It may come as a surprise, but you can't always believe everything you read on the Internet.

Chapters 6, 7, and 8 do not use them, and, the first presented incompleteness theorem in [98] does not use them.

Moreover, as we saw in Chapter 7, Gödel's theorems can be based on natural undecidable *problems*, such as the Halting problem, where the proof we gave does not involve paradoxes or explicit self-referential logic (although many expositions do use such logic to prove the Halting problem).

Undecidable statements, not just problems, were desired Even though natural undecidable *problems* can be used to prove Gödel's incompleteness theorems, since Gödel *sentences* are usually twisty and unnatural, there was for a while a desire to find a natural *undecidable statement* (i.e., a statement that cannot be proven to be true, nor proven to be false in a "standard", consistent logical system of arithmetic). The desire was for a specific statement "which occurs 'naturally' and one which we would reasonably expect to be decidable one way or the other," [50] and yet, was proven to be an undecidable statement.

Lacking such a natural statement supported the view that the only undecidable statements were twisty and unnatural. But from the 1960s onward, several natural undecidable statements were found (see [98, 50] for expositions on one of them), although admittedly, the number is modest.

Concluding Emphasizing twisty, self-referential sentences and paradoxes makes for good press, but the phenomena of incompleteness can occur without them.

9.7.4 Not a Misconception, but a Difference of Opinion

Does Gödel matter? As surprising, deep, and profound as Gödel's incompleteness theorems are, and as forcefully they demonstrate the power of pure reason, most mathematicians are perfectly happy to admire them at a distance and ignore them in their daily work. So, the belief that the phenomena of incompleteness have limited practical impact is understandable. I can't resolve this issue, but I want to emphasize one point.

The phenomena of incompleteness, non-computability, unprovability, and undecidability are highly related (or even equivalent) ways of exposing the inherent limitations of mechanistic, algorithmic, text-based, and rule-based deduction. Gödel showed how this limitation is manifest in axiomatic logical systems; Turing showed how it is manifest in computation; Chaitin showed how it is strongly (alarmingly) manifest in a specific application; and others have shown it manifest in many natural undecidable problems, non-computable functions, and undecidable statements:

> Computability and incompleteness are inherently linked...
> Incompleteness is also famously linked to computability via Chaitin's incompleteness theorem. [60]

Looking at the practical impact of Gödel's incompleteness in isolation ignores the larger universe of related phenomena. And because of the huge and ever-increasing importance of computation, undecidability and non-computability now have more direct practical consequences than does the incompleteness of formal, logical systems. The practical consequences can also extend into pure mathematics even when not explicitly concerned with computation. Mathematical insights on a particular question can often have implications for a related computational problem. But if the computational problem is undecidable, or a function of interest is non-computable, then purely mathematical insights will also be constrained.[41]

The volume of natural undecidable problems (see Section 7.2.4); their use in proving Gödel and Gödel-ish incompleteness (see Section 7.3); the proofs of incompleteness (in Chapters 6, 7, and 8) that don't involve Gödel sentences; the fact that a solution to the Halting problem would lead to a solution of many open questions in mathematics (see Section 7.2); the assertion that the unsolvability of the Halting problem and Gödel's First theorem are *equivalent* (see page 160); and Chaitin's (Gödel on steroids) Theorem 8.3.1, showing an infinity of natural statements that are true but not provable in any consistent logical system; the general applicability of the abstract version of Gödel's First incompleteness theorem (see Theorem 9.5.1), all challenge the view that the phenomena of incompleteness, non-computability, or undecidability are necessarily based on unimportant, unnatural, paradoxical, twisty sentences of the type that Gödel devised. Undecidability can arise in surprising, but natural places.

> ... not only academic problems on Turing machines in computer science can be undecidable, but in fact, natural, physical and intuitive ones as well. [114]

Contrary to the view expressed in [37], the full logical terrain related to Gödel's theorems is large and frequently visited.

> Gödel's unprovable statements ... are not natural mathematical statements: no mathematician has ever stumbled on them (or should we say *over them?*). And thus, it seems to many that normal mathematical practice is not concerned with the incompleteness phenomenon. More and more results show however the contrary [reality]. [60]

[41] For example, the fact that no algorithm that solves all Diophantine equations can exist, will constrain the range of mathematical results one can obtain in the study of Diophantine equations.

9.8 Exercises

1. Is there anything wrong with the following statement about Gödel's First incompleteness theorem? If so, what?

   > ... we know that there must exist statements of arithmetic whose truth we can never confirm or deny. [10]

2. What do you think of the following quote:

   > Gödel showed that there are perfectly reasonable mathematical statements about whole numbers that can be neither proved nor disproved. In a sense, these statements are beyond the reach of logic and arithmetic. [29]

3. Is there anything wrong with the following statement. If so, what?

   > Given any system of axioms that produces no paradoxes, there exist statements about numbers which are true, but which cannot be proved using the given axioms. [37]

4. I recently read an article on impossibility in mathematics. It has the following statement:

   > the logician Kurt Gödel ... proved that it's impossible to prove that everything that is true is true.

   Can you parse that statement? Does it make your head spin? What is it trying to say? If you can parse it, is it correct?

   My answers are *no*; *yes*; *dunno*; *can't parse, so really dunno*. Hopefully you can do better.
5. Berry's paradox is similar to the argument that there are no uninteresting integers. But Berry's paradox is considered real and serious, while the proof that no integers are uninteresting is considered a joke. Can you explain why this is the case?
6. Chaitin's theorem and proof are sometimes claimed to be analogous, or at least related, to Berry's paradox. Try to make explicit any connection you see between Chaitin's theorem and proof, and Berry's paradox.
7. Do you think that proofs of incompleteness that avoid explicit self-reference, Gödel numbers, and Gödel sentences make incompleteness less mystical or divinely inspired?

Appendix

Computer Programs Are Text

This Appendix introduces computer programs for anyone who has never seen one. The only purpose is to explain that a computer program is a *textual document*, which is a fact that is central in all of the impossibility proofs discussed in Chapters 6, 7, 8, and 9. In order to make this point, I will introduce just enough of the programming language Python and develop a few simple programs in Python.[1]

Variables, Values, and Assignments The most basic element of any programming language is the *variable*, which (for our purposes) is just a *name* that can be *assigned* to hold a value (essentially like a variable in algebra). Since a variable can hold a value, we need a way to *assign* a value to it. The statement:

```
variable1 = 0
```

has the effect of *assigning* the value zero to a variable named *variable1*. The name of a variable is chosen by the computer programmer, has no meaning to Python. But it is helpful to choose a name that has meaning to humans. So, the following would be just as good to Python, but would be less informative for any person reading the code.

```
JimmytheGeek  = 0
```

OK, so we can assign values to a variable. How do we use those values? The simplest use is to print out the value a variable is holding. When the following two statements are executed:

```
variable1 = 17
print variable1
```

[1] No, this will not teach you to be a programmer, or even much about Python, but it is enough to make the desired point.

the number 17 will be printed on the computer terminal. Now consider the example:

```
variable2 = variable1 + 3
print variable1
print variable2
```

This takes the current value of variable1 (which is assumed to be 17) and adds 3 to that value, resulting in the value 20, which is then assigned to another variable called *variable2*. Then, the print statement prints the value (17) of *variable1* on one line, followed by the value of *variable2* (20) on the next line.

We can use similar syntax to change the value held by a single variable, as in the following example:

```
variable1 = variable1 + 3
print variable1
```

In this case, 3 is added to the current value of *variable1* (still assumed to be 17), so the result of the addition is 20. Then, that value (20) is assigned to *variable1*. The effect is a change in the value of variable *variable1* from 17 to 20. If you think of the above assignment statement as an algebraic equation, it will make no sense since it uses the same variable on both sides of the "equation". But, if you remember that the statement is an *assignment statement*, where a value on the right-hand side is computed and then that value is assigned to the variable on the left-hand side, it should make sense.

Magical incantation: Input from the user Consider the Python statement:

```
first  = int(raw_input("input the first integer "))
```

I won't try to explain *how* this part of Python works, but *what* it does is print the sentence *input the first integer* on the screen; it next waits for the user to type an integer and press the *Return* key; then it assigns the entered number as the value of the variable named *first*.

Comments In Python, any line that begins with a hash mark (#) is a *comment* that is ignored by Python and is intended to help a human better understand the program. We will use comments in the examples below.

Iteration and Blocks The next construct of Python that we explain is the construct of *blocks*. In order to do that, we must also explain the concept of *iteration*.

One of the most fundamental concepts of any computer programming language is that of *iteration*, that is, the *successive* repeating of some statement(s) in the program. There are several ways to do this in Python. We will only use the main one, which is the *while loop*. Consider the following program:

```
next = int(raw_input("input the starting integer "))
while (next < 25):
      # the following two statements form a block
      next = next + 2
      print next
      # the block ends here

print "Good Bye"
```

The first line asks the user to input an integer. The second line starts a *while loop*. In the *while statement*, the value of variable *next* is compared to the constant number 25. If the value of the variable *next* is less than 25, then the program enters the *block* that follows the *while statement* and executes the statements in the block. If the value of the variable *next* is 25 or greater, then the program does not enter the block.

Note that the *while statement* ends with a colon ":", which indicates the start of a block. A block consists of several statements that are all indented the same amount to the right of the start of the *while statement*. The block ends at the last statement that has that indentation, that is, just before the point in the program where a statement is indented the same amount as the *while statement*.[2] In this example, the block has two statements.

In the block, the value of the variable *next* gets increased by 2; that value is printed out, and then the program returns to the *while statement* to test whether the value of the variable *next* is still less than 25. If it is, then the program enters the block again. So, the effect of the *while statement* is to iterate through the block until the value of variable *next* is 25 or more. At that point, the program skips the block and executes the statement just after the block. That statement prints *Good Bye*, and since nothing follows that statement (and it is not in a block), the program ends.

As an example, suppose the user first inputs the number 30. Then the program will just print *Good Bye* and end. But, if the user inputs the number 18, then the program will print:

[2] This use of indentation to identify blocks is a feature that is almost unique to Python. Some people really love this feature. I don't.

```
20
22
24
26
Good Bye
```

Extending the example a bit In the above program, the value of the variable *next* is compared in the *while statement* to the constant number 25. But, the values of two variables can also be compared in the *while statement*, as in the following program:

```
next1 = int(raw_input("input an integer "))
next2 = int(raw_input("input another integer "))

while (next1 < next2):
      # the following two statements form a block
      next1 = next1 + 2
      print next1
      # the block ends here
print "Good Bye"
```

So, if the user inputs the numbers 17 and 20, the output will be:

```
19
21
Good Bye
```

One last extension Almost all programming languages have *if statements*, where some condition is tested and the result of the test determines what statements are next executed. For example, consider the following modification of the above program:

```
next1 = int(raw_input("input an integer "))
next2 = int(raw_input("input a larger integer "))

while (next1 < next2):
      # this starts an outer block
      next1 = next1 + 2
      print next1

      if (next1 == next2):
         # this starts an inner new block, nested inside of
           the outer block
```

```
            print "The two variables have the same value now,
            which is"
            print next1
            # this is the end of the inner block, which has two
              statements
        # this is the end of the outer block
print "Good Bye"
```

The *if statement* compares the values of variables *next1* and *next2*. If they are equal then an inner block is started, nested inside of the outer block. The inner block has two print statements. When executed, the first print statement prints the sentence inside of the quotes, and the second print statement prints the value of variable *next1*. For example, if the user inputs 10 for the value of *next1* and 18 for the value of *next2*, then the output will be:

```
12
14
16
18
The two variables have the same value now, which is
18
Good Bye
```

You can convince yourself that if input values of the two variables have the same parity (i.e., both even or both odd), then the value of *next1* will become equal to the value of *next2*, and so the print statements in the inner block will be executed. But those print statements will not be executed if the input values do not have the same parity. So, if the two input values were 10 and 17, then the output will be:

```
12
14
16
18
Good Bye
```

Note that in the *if statement* the equals sign (=) is duplicated (==). In Python (and in several other programming languages), when a single equals sign (=) is used, it specifies that an *assignment* should be made, as for example in the statement:

```
variable1 = 0
```

But, when we want to determine whether two variables have the same value, Python syntax specifies that the duplicated equals sign (==) must be used. Using the single equals sign when the duplicated equals sign is required is one of the most common mistakes one makes when programming in Python.

This is enough to make the point OK, we have introduced enough of the programming language Python and written several actual (but rather uninteresting) programs. What you should now understand is that a program is just a *textual document*. Of course, the program must obey specific syntax rules of the language and has a limited number of types of allowed statements. Special programs called *interpreters* or *compilers* can determine whether a text is a legal Python program.

The key point is that (assuming we use a symbol for the end of a line, and include all the spaces used in the program) the lines of a Python program can be concatenated to form a single string; and that string can be re-expanded to recreate the original program exactly as it was written. So, if we start generating all the strings that can be formed using the alphabet of Python (including spaces and a symbol denoting the end of a line), in order of string length (and within the same length, in lexicographic order), then any particular Python program will eventually (but in finite time) be generated. This fact was central and used repeatedly in the impossibility proofs of Chapters 6, 7, 8, and 9.

Bibliography

[1] S. Aaronson. *Quantum Computing Since Democritus*. Cambridge University Press, 2013.

[2] M. Ackerman. *Towards Theoretical Foundations of Clustering*. PhD thesis, Department of Computer Science, University of Waterloo, 2012.

[3] M. Ackerman and S. Ben-David. Measures of clustering quality: A working set of axioms for clustering. *Proceedings of Neural Information Processing Systems 21, Vancouver Canada, 2008*.

[4] K. Arrow. A difficulty in the concept of social welfare. *Journal of Political Economy*, 58:328–346, 1950.

[5] K. Arrow. *Social Choice and Individual Values* (1st edition). Chapman and Hall, 1951.

[6] K. Arrow. *Social Choice and Individual Values* (2nd edition). Yale University Press, 1963.

[7] A. Attiya and F. Ellen. *Impossibility Results for Distributed Computing*. Morgan & Claypool, 2014.

[8] J. Baez. Surprises in logic, 2016. https://math.ucr.edu/home/baez/surprises.html.

[9] P. Ball. *Beyond Weird*. The University of Chicago Press, 2018.

[10] J. D. Barrow. *Impossibility: The Limits of Science and the Science of Limits*. Oxford University Press, 1999.

[11] A. Becker. *What Is Real? The Unfinished Quest for the Meaning of Quantum Physics*. Basic Books, 2018.

[12] M. Beeson. Discussion of incompleteness. Slides for a course on incompleteness and undecidability, lecture 17, Stanford University. www.michaelbeeson.com/teaching/StanfordLogic/Lecture17Slides.pdf.

[13] J. S. Bell. On the Einstein Podolsky Rosen paradox. *Physics*, 1:195–200, 1964.

[14] J. S. Bell. Bertlmann's socks and the nature of reality. *Journal De Physique*, 42:C2–41 C2–62, 1981.

[15] J. S. Bell. *Speakable and Unspeakable in Quantum Mechanics: Collected Papers on Quantum Philosophy*. Cambridge University Press, 1987.

[16] F. Berto. *There's Something about Gödel*. Wiley-Blackwell, 2009.

[17] G. Boolos. A new proof of the Gödel incompleteness theorem. *Notices of the American Mathematical Society*, 36:388–390, 1989.

[18] A. Brown. The weird, but true, evidence for 'spooky action' at a distance (Kavli Hangout). January 8, 2016 www.space.com/31562-weird-universe-revealed-in-quantum-entanglementbreakthrough.html

[19] B. Brubaker. How Bell's theorem proved "spooky action at a distance" is real. *Quanta Magazine*; July 20, 2021; quantamagazine.org.

[20] S. Budiansky. *Journey to the Edge of Reason: The Life of Kurt Gödel*. Oxford University Press, 2021.

[21] A. Calder. Constructive mathematics. *Scientific American*, 146–171, October 1 1979.

[22] G. J. Chaitin. Computational complexity and Gödel's incompleteness theorem. *ACM SIGACT News*, 9:11–12, 1971.

[23] G. J. Chaitin. Gödel's theorem and information. *International Journal of Theoretical Physics*, 22:941–954, 1982.

[24] J. Clauser. Phone interview with John Clauser on the day the 2022 Nobel prizes in physics were announced, 2022. www.youtube.com/watch?v=ZYZiLX2uibM.

[25] J. Clauser, M. Horne, A. Shimony, and R. Holt. Proposed experiment to test local hidden-variable theories. *Physical Review Letters*, 23:880–884, 1969.

[26] P. Coy. The super bowl can teach us about incompatible incentives. *The New York Times*, February 15, 2023.

[27] T. S. Cubitt, M. Gu, A. Perales, D. Pérez-Garcia, and M. M. Wolf. Undecidability in physics: A quantum information perspective, 2022 (in preparation).

[28] T. S. Cubitt, D. Pérez-Garcia, and M. M. Wolf. Undecidability of the spectral gap. *Nature*, 528:207–211, 2015.

[29] T. S. Cubitt, D. Pérez-García, and M. M. Wolf. The unsolvable problem. *Scientific American*, 319:28–37, 2018.

[30] M. Davis and R. Hersh. Hilbert's 10th problem. *Scientific American*, 229:84–91, 1973.

[31] D. Dehlinger and M. Mitchell. Entangled photons, nonlocality, and Bell inequalities in the undergraduate laboratory. *American Journal of Physics*, 70:903–910, 2002.

[32] B. d'Espagnat. Quantum theory and reality. *Scientific American*, 241:158–181, November 1979.

[33] A. Einstein, B. Podolsky, and N. Rosen. Can quantum-mechanical description of physical reality be considered complete? *Physical Review*, 47:777–779, 1935.

[34] J. Eisert, M. P. Muller, and C. Gogolin. Quantum measurement occurrence is undecidable. *Physical Review Letters 108, 260501*, 2012.

[35] A. Ekert. Quantum cryptography based on Bell's theorem. *Physical Review Letters*, 67:661, 1991.

[36] L. Elbroch, M. Levy, M. Lubell, H. Quigley, and A. Caragiulo. Adaptive social strategies in a solitary carnivore. *Science Advances*, 3:e1701218, 2017.

[37] J. Ellenberg. Does Gödel matter?, March 10, 2005. *Slate Magazine*. slate.com/human-interest/2005/03/the-romantic-s-favorite-mathematician.html.

[38] R. A. Fisher. *Statistical Methods and Scientific Inference*. Oliver and Boyd, 1956.

[39] T. Franzen. *Gödel's Theorem: An Incomplete Guide to Is Use and Abuse*. CRC Press, 2005.

[40] S. J. Freedman and J. F. Clauser. Experimental test of local hidden-variable theories. *Physical Review Letters*, 28:938–941, 1972.

[41] J. Geanakoplos. Three brief proofs of Arrow's impossibility theorem. *Economic Theory*, 26:211–215, 2005.

[42] L. Gilder. *The Age of Entanglement*. Vintage, 2009.

[43] R. D. Gill. Statistics, causality and Bell's theorem. *Statistical Science*, 29: 512–528, 2014.

[44] M. Girvan and M. Newman. Community structure in social and biological networks. *Proceedings of the National Academy of Sciences (USA)*, 99:7821–7826, 2002.

[45] K. Gödel. Uber formal unentscheidbare Sätze der Principia Mathematica und Verwandter, Systeme I. *Montshefte für Mathematik und Physics*, 38:173–198, 1931.

[46] D. M. Greenberger, M. Horne, A. Shimony, and A. Zeilinger. Bell's theorem without inequalities. *American Journal of Physics*, 58:1131–1143, 1990.

[47] G. Greenstein. *Quantum Strangeness: Wrestling with Bell's Theorem and the Ultimate Nature of Reality*. MIT Press, 2019.

[48] D. Gusfield. *Integer Linear Programming in Computational and Systems Biology: An Entry-Level Text and Course*. Cambridge University Press, 2019.

[49] L. Hardy. Non-locality for two particles without inequalities for almost all entangled states. *Physical Review Letters*, 71:1665–1668, 1993.

[50] J. Havil. *Impossible? Surprising Solutions to Counterintuitive Conundrums*. Princeton Press, 2008.

[51] D. Hofstadter. *Gödel, Escher and Bach*. Basic Books, 1979.

[52] D. Hofstadter. *I Am a Strange Loop*. Basic Books, 2007.

[53] S. Ing. *Stalin and the Scientists*. Atlantic Monthly Press, 2016.

[54] E. T. Jaynes. Clearing up mysteries: The original goal. In J. Skilling, editor, *Maximum Entropy and Bayesian Methods*, pages 127. Kluwer Academic Publishers, 1989.

[55] E. T. Jaynes. *Probability Theory: The Logic of Science*. Cambridge University Press, 2003.

[56] B. Johannes, T. S. Cubitt, A. Lucia, and D. Pérez-Garcia. Undecidability of the spectral gap in one dimension. *Physical Review X (online)*, 10, 2020. https://journals.aps.org/prx/pdf/10.1103/PhysRevX.10.031038.

[57] J. Kleinberg. An impossibility theorem for clustering. In *Advances in Neural Information Processing, 2002*, pages 463–470, 2002.

[58] A. N. Kolmogorov. Three approaches to the quantitative definition of information. *Problems in Information Transmission*, 1:1–7, 1965.

[59] S. Kritchman and R. Raz. The surprise examination paradox and the second incompleteness theorem. *Notices of the AMS*, 57:1454–1458, 2010.

[60] G. Lafitte. Gödel incompleteness revisited. *Journées Automates Cellulaires*, pages 74–89, 2008.

[61] A. Lewis-Pye and T. Rougharden. Resource pools and the CAP theorem, 2020. arXiv:2006.10698v1.

[62] D. Liben-Nowell. *Connecting Discrete Mathematics and Computer Science*. Cambridge University Press, 2022.

[63] G. Lindsay. *Models of the Mind.* Bloomsbury Sigma, 2021.

[64] R. Lipton. *The P = NP Question and Gödel's Lost Letter.* Springer, 2010.

[65] M. Livio. *The Equation That Could Not Be Solved.* Simon and Schuster, 2005.

[66] S. Lloyd. On the uncomputability of the spectral gap, 2016. arXiv preprint arXiv:1602.05924.

[67] N. Lynch. A hundred impossibility proofs for distributed computing. In *Proceedings of the Eighth Annual ACM Symposium on Principles of Distributed Computing*, pages 1–28, 1989.

[68] N. D. Mermin. Is the moon there when nobody looks? Reality and the quantum theory. *Physics Today*, 38:38–47, 1985.

[69] N. D. Mermin. Quantum mysteries revisited. *American Journal of Physics*, 58:731–734, 1990.

[70] N. D. Mermin. What's wrong with these elements of reality? *Physics Today*, 43:9–11, 1990.

[71] N. D. Mermin. Hidden variables and the two theorems of John Bell. *Reviews of Modern Physics*, 65:38–47, 1993.

[72] N. D. Mermin. Quantum mysteries refined. *American Journal of Physics*, 62:880–887, 1994.

[73] N. D. Mermin. What's wrong with this quantum world? *Physics Today*, 57: 10–11, 2004.

[74] N. D. Mermin. *Quantum Computer Science.* Cambridge University Press, 2007.

[75] C. Moore. Unpredictability and undecidability in dynamical systems. *Physical Review Letters*, 64:2354, 1990.

[76] D. P. Nadlinger et al. Experimental quantum key distribution certified by Bell's theorem. *Nature*, 607:682–686, 2022.

[77] E. Nagel and J. R. Newman. *Gödel's Proof.* New York University Press, 1958.

[78] M. Newman. Modularity and community structure in networks. *Proceedings of the National Academy of Sciences (USA)*, 103:8577–8582, 2006.

[79] M. Nielsen and I. Chuang. *Quantum Computation and Quantum Information.* Cambridge University Press, 2000.

[80] S. Nigro. Why evolutionary theories are unbelievable. *The Linacre Quarterly*, 71, 2004. Available at: http://epublications.marquette.edu/lnq/vol71/iss1/7.

[81] S. Oberhoff. Incompleteness Ex Machina. *Bulletin of the European Association for Theoretical Computer Science*, (128), 2019.

[82] A. Pais. *Subtle Is the Lord: The Science and the Life of Albert Einstein.* Oxford University Press, 1982.

[83] R. Penrose. *The Emperor's New Mind: Concerning Computers, Minds and the Laws of Physics.* Oxford University Press, 1989.

[84] Phys.org. Quantum physics problem proved unsolvable: Gödel and Turing enter quantum physics, December 9, 2015. https://phys.org/news/2015-12-quantum-physics-problem-unsolvable-godel.html.

[85] B. Poonen. Undecidable problems: A sampler. In J. Kennedy, editor, *Interpreting Gödel: Critical Essays*. Cambridge University Press, 2014.

[86] M. B. Pour-El and I. Richards. The wave equation with computable initial data such that its unique solution is not computable. *Advances in Mathematics*, 39:215–239, 1981.

[87] G. K. Pullum. Scooping the loop snooper. www.lel.ed.ac.uk/~gpullum/loopsnoop .html.

[88] P. Reny. Arrow's theorem and the Gibbard–Satterthwaite theorem: A unified approach. *Economics Letters*, 99–105, 2001.

[89] D. Richeson. *Tales of Impossibility: The 2000-Year Quest to Solve the Mathematical Problems of Antiquity*. Princeton University Press, 2019.

[90] R. M. Robinson. Undecidability and nonperiodicity for tilings of the plane. *Inventiones Mathematicae*, 12:177–209, 1971.

[91] B. Rosenblum and F. Kuttner. *Quantum Enigma*. Oxford University Press, 2011.

[92] T. Roughgarden. Foundations of blockchains (lecture 9.6: An impossibility result for proof-of-work protocols), 2022. www.youtube.com/watch?v=8 IwnXURUzUw.

[93] D. V. Roy. On Berry's paradox and non-diagonal constructions. *Complexity*, 4:35–38, 1999.

[94] B. Russell. *The Philosophy of Logical Atomism*. Open Court, 1985.

[95] H. Sanders. Arrow's impossibility theorem of social choice. math.uchicago.edu/ may/VIGRE/VIGRE2007/REUPapers/FINALAPP/Sanders.pdf.

[96] A. Sen. The professor of impossibility. *The Indian Express*, February 27, 2017.

[97] M. Sipser. *Introduction to the Theory of Computation* (3rd edition). Cengage Learning, 2013.

[98] P. Smith. *An Introduction to Gödel's Theorems* (2nd edition). Cambridge University Press, 2013.

[99] P. Smith. *Gödel without (Too Many) Tears*. Logic Matters, 2020.

[100] R. M. Smullyan. *Gödel's Incompleteness Theorems*. Oxford University Press, 1992.

[101] R. M. Smullyan. *The Gödelian Puzzle Book: Puzzles, Paradoxes and Proofs*. Dover, 2013.

[102] D. Styer. *The Strange World of Quantum Mechanics*. Cambridge University Press, 2000.

[103] D. M. Harrison Bell's theorem, 1999. https://faraday.physics.utoronto.ca/PVB/ Harrison/BellsTheorem/BellsTheorem.html.

[104] J. S. Townsend. *A Modern Approach to Quantum Mechanics* (2nd edition). University Science Books, 2012.

[105] R. Tumulka. The assumptions of Bell's proof. In M. Bell and S. Gao, editors, *Quantum Nonlocality and Reality: 50 Years of Bell's Theorem*. Cambridge University Press, 2016.

[106] L. Vaidman. Variations on the theme of the Greenberger, Horne, Zeilinger proof. *Foundations of Physics*, 29:615–630, 1999.

[107] Wikipedia. Conway's game of life, 2023. https://en.wikipedia.org/wiki/Conway %27s_Game_of_Life.

[108] N. Wolchover. How Gödel's proof works. Wired Magazine, July 19, 2020.

[109] N. Yu. A one-shot proof of Arrow's impossibility theorem. *Economic Theory*, 50:523–525, 2012.

[110] W. Zachary. An information flow model for conflict and fission in small groups. *Journal of Anthropological Research*, 33:452–473, 1977.

[111] A. Zeilinger. Quantum teleportation. *Scientific American*, April 2000.

[112] A. Zeilinger. *Dance of the Photons: From Einstein to Quantum Teleportation*. FSG, 2010.

[113] G. Zukov. *The Dancing Wu-Li Masters: An Overview of the New Physics*. Morrow, 1979.

[114] L. Zyga. Classical problem becomes undecidable in a quantum setting, July 9, 2012. https://phys.org/news/2012-07-classical-problem-undecidable-quantum.html.

Index

Printed in the United States
by Baker & Taylor Publisher Services